Daniel Becker

Real-time transport through magnetic quantum dots

Daniel Becker

Real-time transport through magnetic quantum dots

Interplay of decoherence and quantum correlations in nonequilibrium charge transport and real-time spin dynamics

Südwestdeutscher Verlag für Hochschulschriften

Impressum/Imprint (nur für Deutschland/only for Germany)
Bibliografische Information der Deutschen Nationalbibliothek: Die Deutsche Nationalbibliothek verzeichnet diese Publikation in der Deutschen Nationalbibliografie; detaillierte bibliografische Daten sind im Internet über http://dnb.d-nb.de abrufbar.
Alle in diesem Buch genannten Marken und Produktnamen unterliegen warenzeichen-, marken- oder patentrechtlichem Schutz bzw. sind Warenzeichen oder eingetragene Warenzeichen der jeweiligen Inhaber. Die Wiedergabe von Marken, Produktnamen, Gebrauchsnamen, Handelsnamen, Warenbezeichnungen u.s.w. in diesem Werk berechtigt auch ohne besondere Kennzeichnung nicht zu der Annahme, dass solche Namen im Sinne der Warenzeichen- und Markenschutzgesetzgebung als frei zu betrachten wären und daher von jedermann benutzt werden dürften.

Coverbild: www.ingimage.com

Verlag: Südwestdeutscher Verlag für Hochschulschriften GmbH & Co. KG
Heinrich-Böcking-Str. 6-8, 66121 Saarbrücken, Deutschland
Telefon +49 681 37 20 271-1, Telefax +49 681 37 20 271-0
Email: info@svh-verlag.de

Approved by: Hamburg, Universität, Diss., 2011

Herstellung in Deutschland (siehe letzte Seite)
ISBN: 978-3-8381-3267-9

Imprint (only for USA, GB)
Bibliographic information published by the Deutsche Nationalbibliothek: The Deutsche Nationalbibliothek lists this publication in the Deutsche Nationalbibliografie; detailed bibliographic data are available in the Internet at http://dnb.d-nb.de.
Any brand names and product names mentioned in this book are subject to trademark, brand or patent protection and are trademarks or registered trademarks of their respective holders. The use of brand names, product names, common names, trade names, product descriptions etc. even without a particular marking in this works is in no way to be construed to mean that such names may be regarded as unrestricted in respect of trademark and brand protection legislation and could thus be used by anyone.

Cover image: www.ingimage.com

Publisher: Südwestdeutscher Verlag für Hochschulschriften GmbH & Co. KG
Heinrich-Böcking-Str. 6-8, 66121 Saarbrücken, Germany
Phone +49 681 37 20 271-1, Fax +49 681 37 20 271-0
Email: info@svh-verlag.de

Printed in the U.S.A.
Printed in the U.K. by (see last page)
ISBN: 978-3-8381-3267-9

Copyright © 2012 by the author and Südwestdeutscher Verlag für Hochschulschriften GmbH & Co. KG and licensors
All rights reserved. Saarbrücken 2012

Abstract

In this thesis, correlated quantum transport through small quantum dot systems in contact with two metallic, non-magnetic leads is studied. To investigate the interplay of nonequilibrium quantum transport and relaxation (due to the dissipative environment) in this mesoscopic open quantum system, a perturbative diagrammatic approach is used and a numerically exact scheme based on the fermionic path integral is considerably extended.

In the first part of the thesis, the complex conductance signatures of a quantum dot with one spin-split single-particle level are studied in the cotunnelling regime at low temperatures. The regime of weak dot-lead tunnel coupling is tackled with the real-time diagrammatic transport theory. The single-electron Coulomb-blockade valley (Coulomb diamond) can be subdivided into parts differing in at least one of two respects: the kind of tunnelling processes, which (i) determine the single-particle occupations and (ii) mainly contribute to the current. No finite systematic perturbation expansion of the occupations and the current can be found that is valid within the entire Coulomb diamond. Therefore, a non-systematic solution is constructed, which is physically correct and perturbative in the whole cotunnelling regime, while smoothly interpolating between the different regions. With this solution the impact of a spin-flip relaxation on transport is investigated. In particular, the study focuses on peaks in the differential conductance that mark the onset of cotunnelling-mediated sequential transport. It is shown that these peaks are maximally pronounced at a relaxation which is roughly as fast as sequential tunnelling.

The second, main part of this work provides a numerically exact analysis of the stationary charge current and the nonequilibrium quantum dynamics of a spatially fixed spin-1/2 magnetic impurity that is coupled to the

Coulomb-interacting single-level quantum dot. The focus lies on the interplay between the nonequilibrium current and the impurity polarisation in the deep quantum regime, in which all system energies are of the same order of magnitude. First, by adopting it to the minimal magnetic Anderson model, the numerical scheme of the iterative summation of the path integral (ISPI) is reviewed in detail. A generating function for the nonequilibrium current and the orientation of the impurity spin is formulated as a real-time discrete path integral over all paths of (i) the magnetic impurity spin and of (ii) effective Ising-like fluctuating spins due to the Coulomb interaction, which are generated after a Hubbard-Stratonovich transformation. With the use of the ISPI scheme, the sum over all these paths can be carried out numerically, while exactly accounting for all lead-induced self-energies within a sufficiently long, but finite memory time, thereby including all non-Markovian effects on a systematic footing. Extrapolation to vanishing (Trotter) time discretisation and infinitely large coherence times assures, that the results are numerically exact, as long as the scheme converges.

The ISPI scheme is then implemented to provide a systematic study of the stationary charge current and long-time polarisation dynamics of the magnetic Anderson dot. For a wide range of parameters, an initially polarised impurity spin shows an exponentially decaying behaviour (relaxation) due to the interplay of the dot-lead tunnel coupling and the electron-impurity interaction. The numerical results for the current and the spin relaxation rates are compared to a mean-field-type Landauer-Büttiker current and decay rates that are obtained by lowest-order perturbation theory, respectively. As far as the current is concerned, the main effect of the electron-impurity interaction—an increase of the dot's resistivity—can be attributed to the change in the single-electron energy structure and is reproduced qualitatively by the Landauer-Büttiker results with quantitative corrections. This is not the case, however, for the observed correlation effects of Coulomb interaction and electron-impurity scattering or mixtures of both. The perturbative relaxation rates on the other hand only yield a crude, first approximation of the impurity dynamics in the studied, inherently non-perturbative regime in the first place (and only for small electron-impurity interaction, no Coulomb interaction, and rather large temperatures). For the fully interacting dot, the predicted behaviour of the fully coherent ISPI values and the sequential approximation can be diametrically opposed, while the discrepancies between them can be as large as 100%.

Contents

1. **Introduction** — 11
   1.1. Effect of Spin Relaxation on Transport Through a Single-Level Dot — 16
   1.2. Diluted Magnetic Semiconductors — 18
   1.3. Theoretical Approaches for Real-Time Transport — 21
   1.4. Outline — 25

2. **Basic Principles of Transport through a Single-Level Dot with Impurity** — 27
   2.1. The Model System — 28
      2.1.1. The Hamiltonian — 30
      2.1.2. Reducing Parameters by Abstraction: the Infinity Limit — 32
   2.2. Nonequilibrium Transport and Quantum Statistics — 36
      2.2.1. The Open Quantum System (OQS) — 38
      2.2.2. The General Idea Behind the Keldysh Framework — 41
   2.3. Preliminary Remarks to Sequential and Cotunnelling Transport — 51
      2.3.1. Transition Rates and Rate Equations — 52
      2.3.2. Sequential Transport and Coulomb Blockade — 61
      2.3.3. Beyond Sequential Transport: Cotunnelling — 64

3. **Influence of Spin Relaxation on Transport in Cotunneling Regime** — 69
   3.1. Model — 71
   3.2. Master equations — 74

	3.3.	Results	83
	3.4.	Summary	90

4. Iterative Summation of Path Integrals — 93

	4.1.	Generating Function and Fermionic Path Integral	95
		4.1.1. The Short Time Propagator	96
		4.1.2. Fermionic Coherent States	105
		4.1.3. The Gaussian Path Integral	110
		4.1.4. Discrete Hubbard-Stratonovich Transformation	116
		4.1.5. Adding the Remaining Interaction Terms	118
		4.1.6. The Keldysh Partition Function	121
		4.1.7. Adding Source Terms	126
		4.1.8. Tracing over the Electron Degrees of Freedom	128
	4.2.	The ISPI Method	140
		4.2.1. Finite Correlation Length Approach	142
		4.2.2. The Iterative Scheme	148
		4.2.3. Cancelling of Divergent Terms in the Generating Function	151
		4.2.4. Extrapolation Procedures	158
	4.3.	Summary	172

5. Spin-Relaxation by a Charge Current: Exact Results — 175

	5.1.	A Rate Equation for the Impurity Spin	176
	5.2.	Impact of the Interaction on the Charge Current	178
		5.2.1. The Landauer-Büttiker Current	181
	5.3.	First Approximation of Impurity Spin Dynamics: Sequential Flip-Flops	189
	5.4.	Influence of Temperature, Bias Voltage, and Coulomb Interaction	204

    5.5. Summary 212

6. Conclusions and Outlook   215

A. Evolution Operator and Von Neumann Equation   223

B. Sequential Rates: Tracing out the Leads   225

C. The Interaction Picture   227

D. Coherent States, Grassmann Numbers and Gaussian Integration   231

E. Hubbard-Stratonovich Transformation   237

F. Lead Green's Function   241

G. Supplementary Calculations for ISPI   247

H. Landauer-Büttiker Current and Sequential Flip-Flop Rates   259

Danksagung   265

List of Figures

2.1.	The model system	31
2.2.	The open quantum system	38
2.3.	Equilibrium and Keldysh time contours	45
2.4.	Transport scheme with model energies	61
2.5.	Illustration of sequential transport	63
2.6.	Illustration of cotunnelling processes	67
3.1.	Charging diagram of a single-level quantum dot's cotunneling regime	77
3.2.	Current and differential conductance around the inset of cotunneling-mediated sequential transport	84
3.3.	Transport scheme illustrating elastic and inelastic cotunneling in Coulomb-blockade regime	86
4.1.	Illustration of elementary state transitions	99
4.2.	Graphical illustration how to construct the flip-flop polynomials	121
4.3.	Examplary impurity path	135
4.4.	Graphical illustration how to construct the Ξ-matrices	138
4.5.	Exponential decay of inverse Green's function	144
4.6.	Approximate Ξ-matrix for examplary impurity path	147
4.7.	Free Green's function in real-time space	154
4.8.	Segmentation of spin paths	159
4.9.	Extrapolation to vanishing step width of time discretisation	162
4.10.	Extrapolation to infinite coherence time	165
4.11.	Principle of least dependence	169

4.12. Estimation of maximal number of flip-flops per coherence time. 171
5.1. Impurity spin polarisation versus time 178
5.2. Charge current against J 180
5.3. Transport scheme for sweep of interaction strength J . . . 184
5.4. Comparison of ISPI and Landauer-Büttiker current as function of J 187
5.5. Current versus J for four different bias voltages 188
5.6. Impurity spin relaxation time versus J for four different bias voltages 191
5.7. Reducible and irreducible flip-flop diagrams 194
5.8. Dependence of inverse relaxation time on the bias voltage . 201
5.9. Transport scheme illustrating voltage dependence of the decay time 203
5.10. Dependence of relaxation rate on the bias voltage 206
5.11. Dependence of inverse relaxation time on the Coulomb interaction 207
5.12. Current versus Coulomb interaction strength 211

1

Introduction

As one of the greatest achievements of twentieth century physics, the theory of quantum mechanics [1–3] has revolutionized man's way to perceive the physical world and—although its creation dates back almost a century now—is "younger" than ever before and a driving force of scientific and technological progress. A theory of the microscopic world, the realm of entities too small for (direct) human perception, quantum mechanics has always been fascinating for its own sake. It describes a strange, invisible world that is, so to speak, hidden "at the bottom" of the classical world and governed by laws that often are, if not (seemingly) absurd, at least in contradiction to our (common sense) intuitions about nature. It was only decades after the theory was conceived, however, that quantum phenomena could not only be studied and manipulated directly in the laboratory but also found numerous technological applications, such as the laser, the atomic clock, flash memory, and many more. Already the countless places to find a laser application in every day situations illustrates that quantum mechanics has now become an integral part of our lives.

One of its cornerstones is the so-called *particle-wave duality* [4]; in the quantum world, a classical particle acquires a wave-like (dual) nature [5], while in turn classical waves exhibit particle-like behaviour [6]. The wave-like aspects of a physical entity directly entail the existence of characteristic

1. Introduction

phenomena such as the eponymous *quantisation* of a particle's energy, superposition and interference of quantum waves, quantum tunnelling, and entanglement. As it is impossible to experimentally study "pure", i.e., isolated quantum systems[1], it has always been of fundamental interest to connect them to the classical world (bottom-top approach) in ways that allow to see signatures of quantum effects in the non-quantum part of the system (often called *environment*). The H-atom in interaction with an electromagnetic field is a very prominent, if not the paradigmatic example of such an experiment, where the quantization of the valence electron's energy manifests in discrete lines in the luminescence spectrum [7, 8]. Yet, as a trade-off to convenient experimental access to quantum phenomena, the necessity of a connection with an environment entails the statistical effect of decoherence [9–12], which is caused by dephasing [13] and/or dissipation [14–19] and is a limiting factor to coherent quantum dynamics but also responsible for the appearance of classical physics.

In addition to the intentional scientific exploration of the quantum world, recent technological progress in the fabrication of ever-smaller integrated circuits, is about to cross the border between classical electrodynamics and quantum physics (top-bottom). This is both a chance and a challenge. For instance, the isolating oxide layers that separate circuit paths of state-of-the-art microprocessors are only a few nanometres thick. As a consequence, leak currents due to electrons tunnelling or hopping through the isolation grow exponentially with the clock rate and operational voltage [20]. This is clearly an unwanted quantum effect. On the other hand, the ability to produce complex semiconducting and metallic structures of nanometre dimensions allows to integrate quantum systems such as quantum dots (also called artificial atoms) [21–25], molecules [26, 27], carbon nanotubes, sheets of

---

[1] Any measurement requires, at some point, an interaction between the quantum system and the (classical) measuring apparatus.

1. Introduction

graphene (see, e.g., [28] and [29]), and others into conventional semiconductor electronics. The excellent control over materials, methods, and fabrication processes is the reason why semiconductor-based solid state physics, together with quantum optics, is one of the most dynamical, progressive, and comprehensive fields in modern physics. In recent years, steps have been made to integrate optical devices into semiconductor electronics as well, thus combining optics and electronics on a single chip (see [30] for silicon based devices).

Due to a high level of experimental control combined with their structural and spatial simplicity, *quantum dots* are particularly versatile objects of research. They come in many forms: In semiconductor layer systems (e.g., GaAs sandwiched between AlGaAs), metallic top-electrodes can electrostatically "impress" a quantum dot geometry into the two-dimensional electron gas at the GaAs/AlGaAs interface [25, 31, 32] and a similar technique can be applied to quantum wires [33–35]. There are methods to create self-assembled dots [24, 36], for example by depositing materials onto a substrate from the gas phase. Bigger dots and nanoparticles can be made of different kinds of materials (metallic, semiconducting, ferromagnetic, superconducting, etc.), e.g., by wet chemical analysis or by etching them out of semiconductor substrates [37–41]. Additionally, dots can be created within graphene and carbon nanotubes [42–44]. Even a single molecule, trapped between electrodes, can be considered as a quantum dot [45]. This rich variety of materials, sizes, shapes, and structures allows the tailoring of quantum dot systems that meet almost any combination of requirements, which renders them an ideal tool for the exploration of the quantum world.

Over the last decades, quantum dots were therefore heavily employed investigating fundamental quantum phenomena (such as Pauli- and spin blockade [32, 43, 45–49] or the influence of a dissipative environment on the coherent dot dynamics [35, 50–60]) and developing novel devices with ap-

1. Introduction

plications that go beyond those of conventional (classical) electronics. For instance, dots that contain few particles (electrons, holes, excitons) and are connected to macroscopic metallic- or semiconductor leads can be seen as the electronic analogue of the H-atom (or another atom/molecule) in a light field. In transport experiments, the single-particle excitation spectrum of single- and double dot systems manifests in discrete peaks of the differential conductance of a current flowing through the dots via weakly coupled tunnelling leads [31, 32, 47]—analogous to optical spectroscopy experiments. Also, the scope of realised or proposed applications for quantum dots is large. For instance, using them as a *single-electron transistor* [61–64] in an attempt to replace a conventional electronic functional element by a quantum version may seem to be merely aimed at enhancing the efficiency of classical electronics. The mechanisms and physical processes that make it work, however, have an essential quantum nature and are very different from those that apply in a classical transistor. Furthermore, classical electrodynamics often fail to describe systems, for which the electrons intrinsic spin—a genuinely quantum mechanical entity—plays a vital role. This applies particularly to situations when a particle's spin can be addressed and manipulated individually.

The scientific progress with the manipulation of spins and the idea to use them as a fundamental degree of freedom in electronics has spawned a whole new field of research and technology: spintronics [65–69]. As the name suggests (and once they can be realised in experiments), pure spin(-tronic) currents would consist of moving spins rather than charges and corresponding fundamental spintronic devices are spin injectors/pumps and filters/valves, as well as (spin-) detectors, storage capacities (memory), and transistors [70–72]. Compared to conventional electronics, this technology has a lot of (potential) benefits: it may help to build much smaller, more efficient transistors and memory devices that promise to be considerably less

1. Introduction

volatile than charge-based memory (in the ultimate lower limit, individual cells contain a single spin). Furthermore, single-spin storage devices are an ideal candidate for a spin-qubit, the essential ingredient of *quantum computers* [73–75]. A qubit is a quantum mechanical two-state system[2], which can (in contrast to a classical bit) be prepared in a superposition of both the logical 0 and 1 states. Multiple qubits can be entangled, i.e., prepared in a coherent superposition of all possible product states to form a quantum register. *As long as this coherence is preserved*, a logical operation (quantum gate) applied to a register of n qubits is tantamount to applying the operation to all possible 2^n permutations of n classical bits *simultaneously*. A computing machine built of these quantum registers and -gates would allow to exploit this possibility of massively parallel operation. In certain cases, such a quantum computer may be superior to a classical machine, for example, when simulating another quantum system or solving numerically costly, yet partially recursive problems (like the factorisation of a product of big prime numbers). Besides nuclear spins of donors in silicon [76] and Josephson junctions [77, 78], promising candidates to host qubits in solid-state devices are quantum dots [70, 79]. As coherent (unitary) quantum evolution is essential for quantum operations, it is a crucial question how a qubit can be protected from the influence of the (decoherence-inducing) environment.

In the present theoretical work, we ponder on questions in that general, spintronic context, regarding the interplay of coherent quantum dynamics and decoherence effects in an idealised, so-called Anderson model. It consists of a small Coulomb interacting quantum dot (or molecule) with only a single orbital electron-level in tunnelling contact with macroscopic (metallic, non-magnetic) leads as a dissipative environment. While the interac-

---

[2]In case of a spin-1/2, it consists of the up- and down state representing the logical 0 and 1, respectively.

1. Introduction

tion with a single environmental degree of freedom may alter the phase relations of a coherent quantum state in a controlled fashion, it is the aggregated stochastic effect of a macroscopic number of them that, in general, tends to irreversibly destroy the correlations between the different properties of the wave-like quantum state. This is particularly interesting when a bias voltage that is applied between the leads causes a nonequilibrium situation with energy, charge, and spin flowing through the dot—in other words: transport. The model's rather simple structure allows to treat it within a number of different theoretical frameworks, yet it is sufficiently complex to show a rich and non-trivial dynamics. This makes it both a useful object of research for addressing fundamental questions of quantum dynamics and reproducing features seen in experiments with more realistic systems.

1.1. Effect of Spin Relaxation on Transport Through a Single-Level Dot

Even when the coupling is weak, electron- or hole excited states with a definite spin orientation in few electron quantum dots have a finite lifetime due to interactions with various environments (leads, nuclear spin- or phonon bath). A dephasing of the electron spin (transversal relaxation) can be caused, for instance, by the random time-evolution of the effective magnetic field due to the nuclear spins in the dot's host material. A flip of the spin (longitudinal relaxation) can be caused by the dissipation of angular momentum or, in case of non-degenerate spin states, for example by a combined energy- and spin dissipation into both spin- and phonon bath (see below). In experiments spin relaxation is mostly considered as a limiting factor for coherent dynamics, resulting in longitudinal spin coherence times T_1 of anything between μs and hundreds of ms [32, 50, 53, 57, 58, 80, 81],

1. Introduction

depending on quantum dot (its shape, material, structure), temperature and applied magnetic fields [32, 50, 53, 58]. In semiconductor dots, two microscopic mechanisms are considered to cause the flip of the electron spin. The relaxation is induced by acoustic phonons, whose energies match the difference between the spin states, while the spin itself (angular momentum) is absorbed (i) by the nuclear spin bath via the hyperfine interaction or (ii) transferred to orbital degrees of freedom via the spin-orbit interaction [50, 53, 58, 59, 82, 83]. There are cases, however, in which a decoherence effect like spin relaxation can also enhance the visibility of coherent transport signatures. In their paper, Weymann and Barnaś [55] study transport through a single-level dot that is weakly connected to ferromagnetic leads with anti-parallel magnetisation. In this system, a slow spin relaxation can enhance both the zero-bias anomaly of the differential conductance for a spin-degenerate level and the conductance step that appears at the onset of inelastic, coherent second-order tunnelling processes (cotunnelling).

This thesis investigates the effect of an intrinsic spin relaxation on similar transport signatures at the inset of cotunnelling-mediated sequential transport in the Coulomb-blockade regime [84] of a single-level dot with lifted spin-degeneracy. In this regime, purely sequential tunnelling is energetically suppressed due to the mutual Coulomb repulsion between two electrons on the dot. It is only the occurrence of inelastic coherent tunnelling of two (or more) electrons that can populate the single-particle state with the higher energy, which can in turn be depopulated by incoherent processes. This manifests as an increase (step) in the current. We show, that the visibility of the corresponding peak in the differential conductance is enhanced by a small spin relaxation (see also [85]). To that end, we derive a master equation that allows calculating the stationary transport through the dot using the well-established real-time diagrammatic technique by Schoeller and Schön [46]. It is a perturbative method based on the Keldysh technique

1. Introduction

for nonequilibrium dynamics [86, 87] and is particularly apt to evaluate the current close to single-particle resonances [47, 88].

1.2. Diluted Magnetic Semiconductors

In addition to the usual semiconducting and metallic host materials and substrates, another class of materials attracts a lot of interest in the scientific community: *diluted (ferro-)magnetic semiconductors* (for review articles, see Refs. [89–92]). As the name suggests, they combine properties of (ferro-) magnetic materials, e.g., a persistent spontaneous magnetisation below some Curie temperature, with those of semiconductors. They are produced by doping a semiconductor like Ga, ZnSe, CdSe, and others with magnetic atoms like Mn^{2+}, thus embedding spatially fixed magnetic impurities into the substrate (in case of manganese with a spin of $5/2$) [93–96]. Since the impurity's mean relative distance in these materials is rather large (as indicated by the attribute "diluted"), their direct interaction is weak. Itinerant charge- and spin carriers (electrons, holes, excitons), however, can mediated a comparably strong ferromagnetic interaction between the impurities—strong enough to even result in ferromagnetism [97, 98].

As this ferromagnetism is caused by the presence of the itinerant particles, which in turn can be controlled essentially by electrical means, the diluted magnetic semiconductors may allow for an all-electrical manipulation of their magnetisation. This elevates the level of control over spins and magnetism in semiconductor spintronics to a new level. To investigate this phenomenon in quantum dots is particularly promising, for they combine the unique properties of magnetic semiconductors with the quantum nature and versatility of quantum dots. For example, compared to the rather volatile orientation of an electron spin in a dot the spin of magnetic impurities—and even more so a small (ferromagnetic) ensemble of them—

promise to be more stable. At the same time, the magnetic properties of the dot can be switched by tuning the number of itinerant carriers residing on the dot. For very small dots, it was shown that this mechanism essentially depends on the spin of the many-particle state of the itinerant carriers and, therefore, can be sensitive to single-particle changes in the number of carriers [99–103]. In these cases, the total angular momentum of the charge carrier wave function shows a dependence of the particle number that corresponds to Hund's rule for the filling of orbitals in atoms [104]. Abolfath et al. [105] also studied how the carrier-mediated magnetism depends on other dot properties such as the confinement potential, the Coulomb energy, and the temperature. With the ability to change the magnetism in the dot not only via the number of carriers but also by adjusting the confinement potential via a gate voltage, piezomagnetic quantum dots can by produced and used as magnetic switches [106]. Their complex and rich electron-impurity dynamics also allows using magnetic dots as voltage-controlled spin filters [71] and controlling their magnetoresistance via a gate voltage [107].

Experimentally, there has been a lot of research regarding quantum dots with magnetic impurities. Light sources and electromagnetic fields are a very powerful and efficient tool for studying and manipulating these dots [108–114]. In general, optical methods are well-established, reliable, and offer a large "toolbox"[3] for sophisticated experimental manipulation. Most of these experiments, however, have to be performed on ensembles of dots, as it is difficult to address them individually by optical means. Conversely, the fabrication of and experiments with purely electronic systems are generally rather complicated and challenging. On the other hand, electrical means of manipulation using conventional semiconductor electronics allow to study individual dots, while integrating the essential parts of the experi-

---

[3]For example, laser sources that generate ultra-short highly intensive pulses [111].

1. Introduction

ment onto a single chip [115–119]. Due to improvements in the fabrication processes, it is now possible to embed *single* manganese ions into quantum dots and to study their properties [111, 113, 117–119]. Small quantum dots with few (single) carriers and a single magnetic impurity mark the endpoint of the ongoing miniaturisation process in the range of magnetic semiconductors. These systems are certainly interesting to study for their own sake, since they help providing answers to fundamental questions, e.g., concerning the effect dissipation has on coherent quantum dynamics. Furthermore, as the technological development proceeds, small magnetic dots containing single impurity spins may become important candidates for highly efficient, high density spintronic devices and yet another possible solid-state realisation of a qubit. In this context, the role of decoherence is also of vital importance for practical applications.

In the present work, we theoretically investigate the real-time nonequilibrium dynamics of a single-level quantum dot with one fixed, spin-1/2 magnetic impurity. In this simple, idealised version of the magnetic Anderson model, the impurity can only interact with on-dot electrons, which are also subject to the decohering influence of the leads as a dissipative environment. As we are interested in the interplay between the coherent on-dot dynamics and the dissipation due to tunnelling of electrons between the dot and the leads, our studies focus on the deep quantum regime, in which all interactions between system components are of the same order of magnitude (given by some reference energy E). An applied bias voltage between the leads results in a charge current through the dot. While on the dot, the stochastically tunnelling electrons interact with the impurity spin and affect its coherent propagation, which eventually leads to a complete dissipation of an initial polarisation. To properly account for the real-time behaviour of the impurity's dynamics and characterise the dissipation, we have to study long propagation times (compared to $\hbar/E$). To this end, we adopt the method of the *iterative summation of path integrals* (ISPI) that

1. Introduction

was developed by Weiss et al. [120] to the minimal magnetic quantum dot model. In the following section, we compare this method to existing frameworks and explain why it is one of the most suitable choices for describing this kind of model in the deep quantum regime.

1.3. Existing Theoretical Approaches for Real-Time Transport Through Quantum Dots

There are a number of theoretical methods for calculating the real-time dynamics of quantum dot systems. For many decades, various perturbative master equation approaches have been successfully employed to describe transport through quantum dots (see, e.g., Refs. [31, 121]). Besides their generally moderate computational complexity, whenever they provide a high, often excellent empirical adequacy, master equation approaches allow for intuitive explanations of the physics at hand. In the essentially nonperturbative regime considered in the last part of this work, though, these schemes run into principle difficulties, as no small expansion parameter exists. Nevertheless, owing to the mentioned benefits, we employ a real-time diagrammatic technique [46] to obtain a first approximation of the relaxation rate of the polarised impurity spin in the presence of the dot-lead coupling. This is compared with fully-coherent, time dependent numerical ISPI results.

In this context, the scattering Bethe ansatz approach [122–124] deserves a special mention, as it may provide a complete analytical solution of the nonequilibrium, integrable *Anderson dot* model. Aside from remaining technical problems, however, some limitations render it inappropriate to use it for our purposes: At this point, it is restricted to dots with a single, spin-

1. Introduction

degenerate level with on-site Coulomb interaction and can only provide a stationary state solution. A whole class of important analytical methods, which go beyond naive perturbation theory, are based on the *renormalisation group* (RG) [125–134]. Starting from a usual perturbation expansion (e.g., in the dot-lead coupling), flow equations are derived that describe the systematic incorporation of certain (infinite) classes of higher order expansion terms into finite order kinetic equations. These RG methods can be used to analyse non-perturbative transport phenomena like the Kondo effect [126–128] but are restricted to the large bias regime ($VB \gg k_B T_K$), where T_K is the Kondo temperature [135, 136]. Recently, the real-time [125] and frequency formulations [129, 132] of the so-called real-time RG schemes in combination with the functional RG (fRG) proved to be particularly viable to study this regime [129, 130, 133, 134]. Although these RG methods go beyond perturbation theory, the restriction to the large bias regime can be seen as a remnant of their inherent, perturbative foundations. For the case of both small temperature and bias voltage, the Fermi-liquid theory was successfully applied to the Anderson model to obtain the three lowest expansion terms of the self-energy in orders of V_{bias} and T [137, 138]. As of today, cases in which neither the temperature nor the bias voltage is a small parameter of the system (in other words, $V_{\text{bias}} \simeq k_B T_K$) cannot be reliably described by either of these approaches.

In addition to sophisticated analytical methods, a number of numerical approaches have been developed in the last years that allow to simulate nonequilibrium situations in quantum dots coupled to various environments. An important instance is derived from the equilibrium versions of the numerical renormalisation group (NRG) [139, 140]. Based on the introduction of a discrete single-particle scattering basis, it was extended to treat real-time dependent nonequilibrium systems [141–143]. The resulting, powerful and versatile real-time NRG is applicable to finite tempera-

1. Introduction

tures and not restricted to simulating small systems. Due to the fact that, by construction, whole continuous intervals of the energy spectrum are represented by a single, discrete state, however, real-time NRG is not numerically exact. Another scheme that derives from the general RG concept is the real-time version of the density matrix RG (TD-DMRG) [144, 145]. It requires to represent the mesoscopic system at hand, e.g., a quantum dot plus macroscopic reservoirs, by a large but finite lattice-model. For this reason, the propagation times, to which a system can be simulated, are limited by the eventual appearance of finite size effects. As long as transient (short-time) dynamics are considered or whenever the observable of interest (the charge current in most cases) becomes stationary before finite-size effects set-in, the TD-DMRG is a valuable numerical tool.

Furthermore, numerically exact methods based on the quantum Monte-Carlo (QMC) concept have been developed to investigate transport in quantum dots. Their common functional principle consists of finding a (path-) integral formulation of the system's observables and evaluating them via the Monte-Carlo method. The latter draws its effectiveness from a random sampling of the integrand according to a probability distribution that assigns a physical weight to an individual sample (random path). For equilibrium (imaginary-time), the QMC approach is a well-established, powerful tool and possible issues such as the fermionic sign problem are, if not solved, at least controllable. For nonequilibrium (real-time), the situation is more complicated. In addition to the part of the (path-) integrand corresponding to the observable, the function used to measure the physical weight of a random sample is itself highly oscillatory. As a consequence, the summation of a large number of random contributions with opposite sign and comparable absolute value poses severe numerical problems as far as accuracy and computing times are concerned. This dynamical phase problem leads to high computational costs of the existing real-time QMC approaches

1. Introduction

[146–149], effectively restricting the range of accessible parameters. In many cases, for instance, it is difficult reaching, with acceptable stochastic error, propagation times that are long enough to observe the stationary behaviour of the system. On the other hand, despite the fact that these methods were developed only recently, artful optimisations have already been implemented (and can be expected in the future) to stretch the boundaries of rtQMC.[4] Hence, while they are not restricted to small systems or interactions, the existing rt-QMC schemes have problems when long propagation times are needed. A different technique to tackle transport physics with the help of the QMC concept has been devised by Han and Heary [151]. The first step for circumventing the dynamical phase problem is to construct integral representations based on functions that are analytic on the whole complex plane. This is done by the introduction of complex-valued chemical potentials and imaginary "Matsubara voltages" (numerical analytic continuation)[5], which allows to "map" the nonequilibrium (real-time) to an equilibrium (imaginary-time) problem, to which the corresponding standard QMC techniques are employed. The crucial factor here is the finite number of Matsubara voltages used to obtain an approximate representation of the real-time integrand. At present, there remain discussions about the origin of the non-monotonous dependence of the conductance of a Kondo quantum dot system on the bias voltage as reported in Ref. [151] and further, more fundamental questions with regard to the underlying concept of the method. At last, we mention a very recent work of Segal et al. [152], where the authors introduce a numerically exact, iterative path integral simulation method that is very similar to ISPI.

In comparison to these theoretical frameworks, the ISPI scheme [120]

---

[4]For example, Jung et al. [150] are working on an adoption of the dual fermion approach to nonequilibrium.

[5]In analogy to the analytic Fermi function, whose values on the complex plane are given by a discrete (but infinite) sum over Matsubara frequencies

features a different combination of characteristics—one that renders it particularly useful for the purposes of the present work. Similar to the various QMC approaches and contrary the (numerical) ones based on the RG, ISPI is numerically exact and deterministic. This results in rather high computational costs restricting the scope of feasible models to small quantum dots and limits the strength of on-dot interactions to intermediate values. In contrast to QMC, however, the computing times of the ISPI simulations scale linearly with propagation time of the model and, hence, long time dynamics can be studied. Furthermore, ISPI avoids the dynamical phase problem, as the numerical path integration is carried out, for a given timd-descretisation, over all paths in state space rather than random samples of them. At the same time, ISPI accounts for all relevant correlations of the system in the deep quantum regime (provided that the simulation converges for a given parameter set and reasonable running times). The model system we study in this work—a single-level quantum dot with one magnetic impurity—is small enough to be treated with ISPI, while its structure is sufficiently rich to show complex correlated quantum dynamics.

1.4. Outline

The thesis is structured as follows. In chapter 2, the model system is introduced in detail and we explain the essential idealisation steps. It is shown that the model is an instance of an (mesoscopic) *open quantum system*, an important concept that is outlined shortly. The rest of the chapter is devoted to the Keldysh framework as a necessary means to describe transport beyond linear response. It also features preliminary remarks to the notion of transport itself, illustrated by basic phenomena caused by sequential tunnelling and cotunnelling in the regime of weak dot-lead coupling. Building on that, chapter 3 contains an investigation of the influence of electron

1. Introduction

spin relaxation on the appearance of conductance peaks that mark the inset of cotunnelling-mediated sequential transport in the Coulomb-blockade regime. The analytical calculations are based on perturbative rate equations, which are obtained with the real-time diagrammatic technique by Schoeller and Schön [46].

In chapter 4, we adopt the ISPI approach to the single-level Anderson dot with a spin-1/2 magnetic impurity. After deriving a path integral representation of the Keldysh partition function, a formally exact generating function is constructed by adding proper source terms to the action. The subsequently described, central element of the ISPI scheme allows to systematically cut irrelevant time correlations beyond some finite memory time and to calculate the generating function iteratively. We explain how the numerical exactness of the method is preserved via appropriate extrapolation schemes that allow to eliminate systematic errors due to time discretisation and the finite memory. The introduction of the method itself follows a systematic analysis of the impurity spin dynamics and stationary charge current in chapter 5 based on ISPI. We focus on the relaxation (dissipation) of a finite impurity spin polarisation in presence of the electron tunnelling between dot and leads and study the dependence of the corresponding relaxation rate on several combinations of model parameters. These findings are compared to perturbative results obtained for a relaxation rate that is caused solely by sequential electron-impurity spin flips (flip-flops). We also present a systematic analysis of the current, which is compared to an approximate Landauer-Büttiker result. Chapter 6 both summarises the results of the thesis and provides an outlook to (i) possible future research and (ii) further applications of the ISPI method. We provide supplemental calculations and remarks in appendices A-H.

2

Basic Principles of Transport through a Single-Level Dot with Impurity

THE CENTRAL THEME of this thesis is the quantum mechanical transport of particles, charge and spin through a nanoscale device in contact with a macroscale environment. The interplay of different physical theories—in particular of quantum dynamics and statistical physics—, the involvement of such different concepts as electron tunnelling on one side and quantum dissipation and decoherence on the other side, acting together in a single system, impose particular theoretical challenges, while offering insight into rich and complex physics. When dealing with quantum transport, much about the phenomena that govern the mesoscopic world can already be learnt from the most elementary and idealised model systems.

For our work, we choose the single-level quantum dot (SLQD) as object of research. Of all nanoscale quantum systems, a zero-dimensional quantum dot with a single electron orbital is certainly among the very simplest. It is due to its lack of internal structure, that this basic model grants a comparably undisguised view on the fundamental physics involved. Over the last years, SLQDs proved to be a valuable instrument for theoretical *and*

2. Basic Principles of Transport through a Single-Level Dot with Impurity

experimental investigations of transport in nanostructures (see, e.g., [68]). And although this elementary model does not, of course, represent a realistic quantum dot in all its details, its structural resemblance is in many cases adequate to reproduce—at least qualitatively—the experimental results. Aside from conceptual arguments, a model as simple as the SLQD is frugal in its demands for computational power, be it analytical or numerical. Its usage stands in the tradition of bottom-up physical research to start exploring new physics from the smallest set of assumptions that is sufficient to reproduce empirical data or predict novel effects. The less structure and complexity is needed for an adequate representation of the physics at hand, the higher is the conceptional and computational efficiency.

In this chapter, we present the model in detail and prepare the ground for the more technical considerations in the later parts of the thesis, by giving a short introduction to the basic concepts connected to nonequilibrium and transport in a mesoscopic system. Particularly, we will outline one of the standard ways how mesoscopic transport and quantum statistics are connected using the concept of the *open quantum system* within the framework of the nonequilibrium *Keldysh technique*. We close the chapter with introductory remarks on sequential transport in the SLQD.

2.1. The Model System

The idealised system that will be considered in this work consists of a nanoscale central region (CR) coupled to two metallic leads L and R via identical tunnelling barriers. An additional gate electrode is used to tune the electrostatic dot potential. Figure 2.1 shows a schematic picture of the system. A bias voltage V_{bias} will be symmetrically applied between the leads, which act as electron reservoirs, and shifts their electrochemical potentials relatively to each other. In this simple model, it is assumed that the voltage

2. Basic Principles of Transport through a Single-Level Dot with Impurity

drops instantaneously at the tunnelling barriers and is piecewise constant in the respective regions. Depending on the internal structure of CR and the system parameters, this can cause a charge current of electrons, which tunnel through the barriers. Inspired by realistic experimental set-ups, it is assumed, that the leads are much bigger than the small, point-like central region and can therefore be described as gases of free electrons [153]. We will go into more detail on the specific features of the leads' model representation in the following paragraphs (in particular, sections 2.1.2 and 2.2.1).

We already stated above that the central region we will be dealing with is an idealised electron quantum dot with a single electronic orbital. Besides the orbital structure, the simplest model we take into account—in chapter 3—at least includes the electron spin as well. The basis of electronic eigenstates of the isolated dot therefore has four elements $|\chi\rangle$, where $\chi \in \{0, \uparrow, \downarrow, d\}$, and is characterised in general by the three scalar parameters $\bar{\epsilon}$, Δ, and U. Symbol χ denotes, whether the dot is empty (0), contains one electron with spin $\sigma = \uparrow, \downarrow$, or is in a spin singlet state with double occupation (d). With the single-electron energies ϵ_σ, we define the mean energy $\bar{\epsilon} := (\epsilon_\uparrow + \epsilon_\downarrow)/2$ and the difference $\Delta := |\epsilon_\uparrow - \epsilon_\downarrow|$. An applied gate voltage V_gate can be used to adjust the dot's electrostatic potential $\Phi_D := e\alpha V_\text{gate} + \Phi_0$ and, thus, to shift $\bar{\epsilon}$. Here, $e < 0$ is the elementary charge, α some proportionality constant, and Φ_0 an arbitrary energy offset. Hence, by choosing an appropriate Φ_0, the mean energy can be identified with dot potential: $\bar{\epsilon} = \Phi_D$. If the energy ϵ_0 of the empty dot is set to zero, we achieve this by defining zero potential $\Phi_D = 0$ as the point, at which the (spin-)average dot energy does not change upon charging the empty dot with one electron. Without specifying a particular physical origin, we allow for a non-zero level splitting Δ; the reason could be an external magnetic field or an intrinsic splitting due to crystallographic effects. Lastly, the scalar energy parameter $U > 0$ characterises the Coulomb repulsion,

2. Basic Principles of Transport through a Single-Level Dot with Impurity

which is "felt" by two electrons in the small dot's confinement. Hence, the energy of the two-electron state is given by $\epsilon_d = 2\Phi_D + U$.

In chapter 4, we embed a *spatially fixed* magnetic impurity into the quantum dot, which is represented by a quantum mechanical spin $1/2$. It can interact with a *single* electron in the dot by the magnetic interaction $4J/\hbar^2 \hat{M} \cdot \hat{S}$, where $\hat{M}$ and $\hat{S}$ are the spin operators of the impurity and the electron, respectively. Since two electrons in the SLQD form a singlet state $|d\rangle$ with zero total spin, they do not interact with M. Similar to the one-electron states of the dot, we allow for a non-zero splitting Δ_{imp} of the impurity eigenstate energies $|\tau = \uparrow, \downarrow\rangle$.

2.1.1. The Hamiltonian

The Hamiltonian $\hat{H}$ of the entire system is the sum of the individual parts $\hat{H}_{\text{dot}}$, $\hat{H}_{\text{leads}}$, and $\hat{H}_T$ representing the dot, leads, and tunnel coupling, respectively. They are given by

$$\hat{H}_{\text{dot}} = \hat{H}_{\text{dot}}^{\text{el}} + \hat{H}_{\text{imp}} + \hat{H}_{\text{int}} \quad \text{— impurity part} \tag{2.1a}$$

$$\hat{H}_{\text{imp}} = \Delta_{\text{imp}} \hat{\tau}_z / 2 \quad \text{(only chapter 4 and following)} \tag{2.1b}$$

$$\hat{H}_{\text{int}} = 4J/\hbar^2 \hat{M} \cdot \hat{S} \equiv J \sum_\sigma (\sigma \hat{\tau}_z \hat{n}_\sigma + \hat{\tau}_\sigma \hat{d}_{-\sigma}^\dagger \hat{d}_\sigma / 2) \tag{2.1c}$$

$$\hat{H}_{\text{dot}}^{\text{el}} = \sum_\sigma \epsilon_\sigma \hat{n}_\sigma + U \hat{n}_\uparrow \hat{n}_\downarrow \equiv \sum_\chi \epsilon_\chi |\chi\rangle\langle\chi| \tag{2.1d}$$

$$\hat{H}_{\text{leads}} = \sum_p \hat{H}_p \quad \text{with} \quad \hat{H}_p = \sum_{k\sigma} \epsilon_k \hat{n}_{k\sigma p} \tag{2.1e}$$

$$\hat{H}_T = \sum_{k\sigma p} \gamma \hat{d}_\sigma^\dagger \hat{c}_{k\sigma p} + \gamma^* \hat{c}_{k\sigma p}^\dagger \hat{d}_\sigma, \tag{2.1f}$$

where $\hat{n}_\varkappa = \hat{f}_\varkappa^\dagger \hat{f}_\varkappa$ is the particle number operator for fermions f in a state with index $\varkappa$. For dot electrons and $\varkappa = \sigma$, we have $f = d$, whereas

2. Basic Principles of Transport through a Single-Level Dot with Impurity

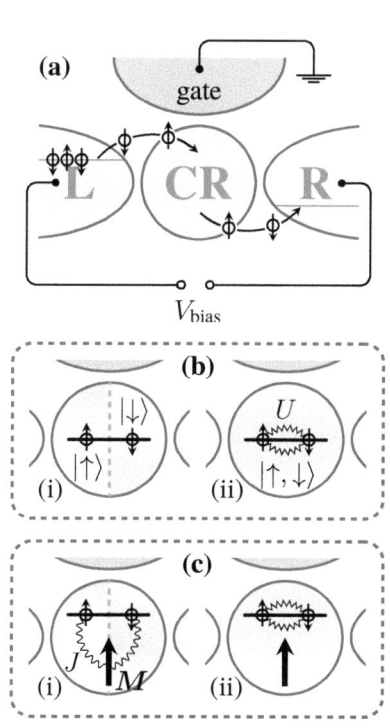

Figure 2.1.: Schematic picture of the two model systems, considered in this work. **(a)** General structure of both models, which differ only in the structure of the nanoscale central region (CR). Two metallic, macroscopic leads (L and R) are tunnel coupled to CR via identical barriers and serve as electron reservoirs. With the gate electrode, the electrostatic potential is adjusted. An applied bias voltage V_{bias} between L and R (horizontal lines symbolize potential levels) leads to a current of tunnelling electrons (small circles with arrows) **(b)** In the simplest case, the CR is a quantum dot with a single orbital electron level (indicated by horizontal black line), which can be empty or occupied *either* by (i) one electron in spin up $|\uparrow\rangle$ or down $|\downarrow\rangle$ state *or* (ii) two electrons in a Coulomb-interacting singlet state $|\uparrow,\downarrow\rangle$. U gives the strength of repulsion. **(c)** In chapter 4, we add a fixed magnetic impurity M with spin 1/2 to the dot. (i) A *single* electron on the dot interacts with M via a spin-spin interaction (characterized by J), while (ii) for double occupation only Coulomb interaction is present.

for the leads and $\varkappa = \{\boldsymbol{k}, \sigma, p\}$, we have $f = c$. The $\hat{d}_\sigma$ and $\hat{c}_{\boldsymbol{k}\sigma p}$ are the corresponding annihilation operators, σ denotes the electron spin, $\boldsymbol{k}$ is a lead electron's wave vector, and $p \in \{\mathrm{L}, \mathrm{R}\}$ the lead index. With the electrostatic potential $-\mu_p$ of lead p due to bias voltage V_{bias}, the energy of an electron with wave vector $\boldsymbol{k}$ is given by $\epsilon_{\boldsymbol{k}p} = \epsilon_{\boldsymbol{k}} - \mu_p$. The tunnelling Hamiltonian $\hat{H}_T$ is responsible for a transfer of electrons between dot and leads, where the first term on the r.h.s. of (2.1f) describes tunnelling from a lead onto the dot, while the second term describes the reverse process. The spin of the tunnelling electron is conserved [154].

The operator for the impurity was identified with $\hat{\boldsymbol{M}} = \hbar/2(\hat{\tau}_x, \hat{\tau}_y, \hat{\tau}_z)$, where the $\hat{\tau}_\varkappa$, with $\varkappa \in \{x, y, z, +, -\}$, are the corresponding Pauli matrices; we use the common convention $\hat{\tau}_\pm = \hat{\tau}_x \pm i\hat{\tau}_y$. An analogous representation can be employed for the dot electron's spin operator with corresponding matrices $\hat{\sigma}_\varkappa$. To arrive at homogeneous expressions for all parts of the Hamiltonian, however, we replaced all instances of $\hat{\sigma}_\varkappa$ by the equivalent second quantised forms, using $\hat{\sigma}_z = \hat{n}_\uparrow - \hat{n}_\downarrow$ and $\hat{\sigma}_+ = \hat{d}_\uparrow^\dagger \hat{d}_\downarrow$. Wherever in this thesis one of the indices σ, τ or p are used as numbers [as in (2.1c), for instance], they are interpreted as signs, where $\uparrow$ and L correspond to $+1$, as $\downarrow$ and R correspond to -1. On this note, we define $\mu_p = peV_{\text{bias}}/2$. Since the structure of both leads is identical on the Hamiltonian level, both μ_p have to be equal in the zero-bias point. For simplicity, we can then set these coinciding zero-bias potential values to zero: $\mu_p(V_{\text{bias}} = 0) = 0$. After we introduce the equilibrium approximation in section 2.2.1, we will be able to identify the μ_p with the *electrochemical potentials* of thermal leads.

2.1.2. Reducing Parameters by Abstraction: the Infinity Limit

We want to complete this section with some remarks about the intended interpretation of the Hamiltonians $\hat{H}_{\text{leads}}$ and $\hat{H}_T$ in equations (2.1e) and (2.1f), respectively. Our goal is to study the interplay between and the basic mechanisms of nanoscopic transport with the help of an abstract, idealised model that has as little structure and few parameters as needed to still represent the scrutinised class of real physical systems adequately. Aside from a high computational efficiency, what is gained from this approach is a relatively clear view on the physics. The less parameters an adequate model has, the easier may their mutual dependencies be accessible by intuitive ex-

2. Basic Principles of Transport through a Single-Level Dot with Impurity

planations. There are many ways of how certain kinds of "non-essential" parameters may be eliminated from a model representation.

As a very common example, which is applied to our model (several times), we mention *infinity limit idealisations* and illustrate them shortly by means of the tunnelling Hamiltonian $\hat{H}_T$. It has a very simple form: the tunnelling is described by just a single constant parameter γ—the tunnelling amplitude. This means, that the coupling $\hat{H}_T$ connects the dot states *with all lead states* equally strongly, i.e., independent of the energies of the states involved. This is a strong simplification, unphysical at the latest when their number is infinite and γ is non-zero. There are ways to mitigate this situation. We could choose a finite system, normalise $\hat{H}_T$ with the a suitable parameter such as, e.g., the number of particles $\mathcal{N}$, but would then have to drag an insignificant parameter through all calculations. Or we could introduce an appropriate dependence on the lead states ($\gamma \to \gamma_{k\sigma p}$, for example an energy cut-off), eliminate $\mathcal{N}$ by setting it to infinity, and pay the price of an infinite number of new parameters (which are disposed of once again at a later stage of the calculations). This path is, in fact, often followed, since it avoids serious mathematical problems due to involvement of low lying energy states that may arise for a Hamiltonian like $\hat{H}_T$. These problems originate from the fact, that in contrast to the infinite lead size, which is merely a counterfactual property, the structure of the coupling term is even manifestly unphysical.[1] Still, we are going to adhere to $\hat{H}_T$ in the present form. We argue that with a proper interpretation of expression (2.1f) and some care applied to the transport calculations, the conceptional and mathematical issues can be resolved—at least, for our particular system. The benefit is, again, that we can start our treatment with the intended, minimised set of parameters. Similar to the lead Hamiltonian, the coupling term $\hat{H}_T$ has to be understood as a part of an abstract model, which repre-

---

[1] For $\mathcal{N} \to \infty$ and finite γ, $\hat{H}_T$ violates the conservation of probability [155].

2. Basic Principles of Transport through a Single-Level Dot with Impurity

sents a certain class of (more) realistic physical systems that is both of finite size and has state dependent amplitudes $\gamma_{k\sigma p}$.

To see which conditions a system has to meet in order to fall into this class, we consider the size $\mathcal{E}_{\text{tun}}$ of the energy interval that is bounded by the minimal and maximal energy differences, so-called *transport channels* (for a more precise definition, see section 2.3.1), between dot states of allowed single-particle tunnelling processes. For our purposes, it can be roughly approximated by the sum of characteristic dot energies and defines the dominant energy scale for tunnelling. That is to say, $\mathcal{E}_{\text{tun}}$ gives the order of magnitude, by which energies of lead electrons that mainly contribute to transport differ. This is due to the conservation of energy, which causes a rapid decrease of tunnelling transport involving reservoir electrons with energies far away from any of the dot's transport channels. In the case of model Hamiltonian $\hat{H}_{\text{dot}}^{\text{el}}$, it amounts to $\mathcal{E}_{\text{tun}} = \Delta + U$ (see section 2.3.1). Which lead electrons actually are involved in the transport, also depends on the bias and gate voltage, or rather the associated energies eV_{bias} and Φ_D, respectively. If we further define $\mathcal{E}_{\text{var}}$ as the energy scale, on which the amplitudes $\gamma_{k\sigma p}$ and the density of states $\varrho(K)$ change their absolute value considerably, we arrive at at least one necessary conditions for a system to be representable by $\hat{H}_T$: all energy windows that are relevant for transport have to be much smaller than the scale on which the tunnelling amplitude changes, i.e., $\mathcal{E}_{\text{var}} \gg \max(\mathcal{E}_{\text{tun}}, eV_{\text{bias}}, \Phi_D)$. This condition ensures, that the absolute value of the $\gamma_{k\sigma p}$ is basically constant for all states with wave vector k that lie in the energy interval of electrons contributing to transport.

We want to stress, that this is not a sufficient condition to avoid the mathematical problems mentioned above. Rather, this has to be tried and checked for every system and approach that is used to describe transport. As we will see in the next chapters, for model (2.1) and both the perturbative approach developed by Schoeller and Schön [46] and the ISPI approach by Weiss et al.

2. Basic Principles of Transport through a Single-Level Dot with Impurity

[120], those difficulties do not arise or can be circumvented. In the spirit of this section, this can be interpreted as the infinity limit $\mathcal{E}_{\text{var}} \to \infty$. A spin and lead dependence of γ was excluded already in the beginning, where we specified the leads as unpolarised and the tunnelling barriers as identical. But even if the absolute value of γ is constant, that would still leave the possibility that γ depends on the phase of a tunnelling electron. In view of the system at hand, however, it is safe to assume that the phase of electrons approaching the tunnelling junctions from the depth of the macroscopic reservoirs is a random function with respect to the direction of k. Based on this supposition, it can be deduced that the tunnelling amplitude would be independent of an electron's incident angle even for general systems (see, e.g., Refs. [88, 156]).

Here, we do not recapitulate the argumentation in detail, as it is not needed in case of our simple model system: In the single-level dot and with the given tunnelling Hamiltonian, an electron that tunnels into or out off the leads is *uniquely* associated with a transition between two of the dot's four different eigenstates $|\chi\rangle \in \{0, \sigma, d\}$. If, for instance, the dot is empty, a spin-up electron that tunnels onto the dot necessarily transfers it into eigenstate $|\uparrow\rangle$. For dots with two or more single-particle orbitals, whose (interacting) eigenstates are coherent superpositions of the single-particle states, there are in general two or more possible paths (in state space) for tunnelling between a given pair of dot eigenstates. Only in such systems might a relative, non-zero phase, associated with different tunnelling amplitudes, be observable. It is in these cases, that the assumed randomness of this relative phase (with respect to the propagation direction of the tunnelling electrons) leads to a destructive interference of coherent contributions due to a superposition of different tunnelling paths. For the simple model that is studied in this thesis, however, such superpositions cannot be created in the first place.

2.2. Nonequilibrium Transport and Quantum Statistics

This section delivers a short introduction of the basic assumptions and ideas that are used to integrate the theoretical concepts of nonequilibrium and transport phenomena with the fundamental theory of statistical quantum physics. In this context, the notion of the *partition function* and of the more general *generating function* will play a major role in the theoretical considerations throughout this work.

The partition function Z is a central and fundamental concept both in thermodynamics and statistical physics. It contains in compact form all the statistical information that allows to describe an infinitely large system by few aggregated properties, as long as its microscopic state distribution is characterised by a thermodynamic (equilibrium) ensemble. Here, the infinite system is to be understood as an idealisation of a finite, real, yet "very large" system, in which the statistical effects dominate the physical behaviour. Often, the word "macroscopic" is used synonymously for "very large" in a thermodynamic context and a prominent example is a cubic centimetre of air with its $\sim 10^{20}$ atoms. Depending on the particular physical situation, however, it can be adequate to already consider significantly smaller systems as infinitely large. Equilibrium states are reached only after infinite ("very long") propagation time and are characterised—according to the second law of thermodynamics—by a stationary value of the entropy S. The infinity in time has, as the one in size, to be understood as a feature of the underlying idealised concept (see section 2.1.2).

Due to the huge number of degrees of freedom, a full dynamical description of each and every microscopic component of a macroscopic system is not useful or even possible. In these cases, a statistical (thermodynamic)

2. Basic Principles of Transport through a Single-Level Dot with Impurity

approach is inevitable, since it allows to reduce the number of relevant parameters drastically to a set of few statistical state quantities, such as an (equilibrium) temperature T, pressure p, and chemical potential μ. Statistical terms can then be calculated with the use of Z as generating function. If Z is the partition function of a grand canonical ensemble, the expectation value of the total number of particles, for example, is given by

$$N = \partial_\mu \ln Z/\beta, \qquad (2.2)$$

where $\beta^{-1} = k_B T$ and k_B is the Boltzmann constant. Hence, a thermodynamic representation combines a high empiric adequacy with a largely reduced descriptive complexity in terms of intuitive and (experimentally) controllable quantities (such as the temperature). But, at the same time, its validity is strictly limited to thermal equilibrium. In consequence, thermodynamics is not (directly) applicable to any nonequilibrium regime beyond linear response (see, e.g., Mahan [157]), where transport properties are connected to equilibrium quantities via the *fluctuation dissipation theorem* [158, 159].

This immediately raises the question, how to deal with coherent transport through a microscopic quantum dot that is interacting with macroscopic leads. Transport and real-time dependence imply a nonequilibrium situation, while macroscopic leads with applied bias voltage (electrochemical potential) and temperature should be described (locally) in thermodynamical, i.e., equilibrium terms. But this is only a seeming contradiction, since statistical phenomena such as decoherence and dissipation are essential to describe realistic transport of, e.g., directed currents beyond coherent oscillating dynamics of (small) closed systems. It can often be resolved by combining a proper nonequilibrium framework with a model representation that "contains" as much thermodynamics as possible. The concept of

2. Basic Principles of Transport through a Single-Level Dot with Impurity

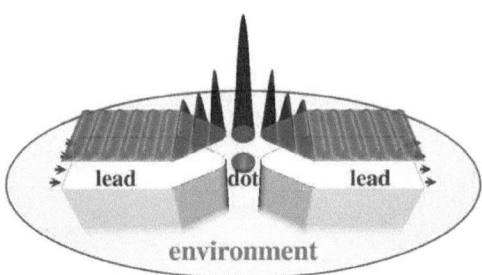

Figure 2.2.: Schematic illustration of our model (leads and dot) as an open quantum system (OQS) embedded in an (unspecified) environment. The arrows denote a directed effective (particle) current through the OQS. The whole system (OQS plus environment) is closed but in nonequilibrium. The wave-like patterns symbolise the flow through the system. Their local relative amplitude in the leads is very small, both due to the macroscopic dimensions of the leads themselves and the high resistivity of the nanoscopic dot, which acts as a bottleneck for interlead particle transfer. If the rate of transferred particles is much smaller than $1/\tau_{eq}$, with lead equilibration time τ_{eq}, it is reasonable to assume that each reservoir stays in its individual equilibrium state. Compared to the microscopic size of the dot (hemisphere), however, the resulting concentrated particle flow (central peak) is large and deviations from equilibrium cannot be neglected.

the *open quantum system* is a well-established framework for such a representation. It allows us to build our nonequilibrium approach on top of an (adopted) notion of the partition function. In the following, we shortly discuss the open quantum system and proceed with an introduction to the nonequilibrium framework named after L.V. Keldysh [86, 87].

2.2.1. The Open Quantum System (OQS)

As the name implies, the essential feature of an open quantum system [160–162] is that it is not closed, i.e, one or more of its extensive quantities such

2. Basic Principles of Transport through a Single-Level Dot with Impurity

as energy, particle number, or others may be interchanged with an unspecified environment. Only the combination of the OQS with this environment is assumed to be closed. It is therefore crucial that it is not part of the description of the OQS itself, hence, "invisible" and not explicitly modelled. Conceptually, this situation resembles, e.g., the underlying model for the grand canonical statistical ensemble of thermodynamics. There, a small fraction of a closed system can exchange energy and particles with the much larger remainder, which in turn acts as a reservoir or environment. But in contrast to any thermodynamic model, a general OQS may be (and often is) in a nonequilibrium state.

Figure 2.2 shows a sketch of a nonequilibrium OQS similar to the model used in this work and of the environment it is embedded in. The open system itself consists of two "very large" (infinite) leads, which are connected to each other via a quantum dot or another nanoscale constriction. A directed total current of particles, energy or other quantities through the leads and the dot is indicated both by small arrows at the interface between OQS and environment and wave-like patterns. Due to the largely differing proportions of leads and dot, the resistivity of the whole OQS for any kind of current flow is determined solely by the dot. The same current that already yields strong deviations from the dot's thermal state, will cause only negligible fluctuations in the macroscopic leads. This is symbolised by the local relative amplitude of the wave patterns compared to the system size. The small curling waves in the leads nearly vanish in view of their spatial dimensions. In the dot, however, the same waves are not only relatively large compared to its size but also grow in amplitude, when squeezed through the nanoscale constriction.

In such situations, it may therefore be reasonable to assume that each lead is always in a thermodynamic equilibrium state. All nonequilibrium physics and time dependence is then contained in the dynamics of the dot

2. Basic Principles of Transport through a Single-Level Dot with Impurity

and maybe its immediate vicinity. This is tantamount to the assumption, that the average (waiting) time τ_T between two tunnelling events is very long compared to the lead equilibration time τ_{eq}: $\tau_{eq} \ll \tau_T$. In this thesis, we consider a theoretical model of a quantum dot with just one orbital level and assume that such an approximation applies. The decision, whether or not a particular material or theoretical system can be successfully mapped onto our idealised model, depends on the individual case and the considered transport regime.

Although it is not itself part of the model, the *environment*[2] plays an important implicit role in the concept of the OQS. For one thing, it contains the unspecified source that provides an everlasting nonequilibrium situation. In the resulting idealised OQS model, this manifests itself in two equilibrium leads with differing electrochemical potentials, which—despite of being in (very high resistivity) contact—never reach some common equilibrium state with chemical potential $(\mu_L + \mu_R)/2$. The particular structures of the current source and potentially complex external dynamics are, at the same time, kept out of any consideration. Only in the vicinity of the contact point, viz., the quantum dot, the externally caused nonequilibrium "re-appears." Finally, due to this "invisible" environment, the OQS has no boundaries at which particles could be (back-) scattered. In the case of figure 2.2, this means that the right-moving particles, once they passed the constriction, will not be reflected at any boundary and never return to interfere with the dot. The environment provides for an infinite return time of particles that move away from the dot or the central region, in general. This is a particularly useful property, when the the main focus of attention lies on the stationary or long-time nonequilibrium dynamics [123, 124].

---

[2] Also called *super-environment*, sometimes, to clearly distinguish it from the leads as part of the system.

2.2.2. The General Idea Behind the Keldysh Framework

With the open quantum system, we now have a model representation at hand, most parts of which are described by thermodynamics. Thus, the remaining task is to find a way to combine these local equilibrium parts with the nonequilibrium dynamics of the dot. The two different approaches we use in this work are both based on the prevalent *Keldysh technique* [87], which has proved itself to be versatile and adaptable to a wide range of models. A comprehensive, yet concise review can be found in the work of Kamenev and Levchenko [86]. The following remarks base on that paper and provide a short introduction to the general idea behind the Keldysh framework. Details of how it is applied to our model are given in chapter 3 and section 4.1.

Initial State and Stationary Transport

All transport theories, that are capable of describing time dependent (as opposed to stationary) quantum dynamics, have to refer to an initial configuration at some given time t_i, generally described by a density matrix $\hat{\rho}_i$.[3] This is simply owed to fact that the Schrödinger-equation is a differential equation in time. When—as in chapter 3—only stationary dynamics is of interest, however, this reference might at first glance seem expendable or even unwanted, as it is the case, e.g., for particular master- and rate equation approaches (see chapter 3). Here, the equations of motion for system states are transformed into self-consistent linear equations by going to the *stationary limit*. The motivation is to find a state, into which the system will propagate after an infinitely long time and *regardless of its initial state*.

---

[3] Throughout this thesis, we will often use the word *state* as referring to a density matrix (operator) rather than to a Hilbert space vector $|\psi\rangle$, which can be represented by the corresponding projection operator $|\psi\rangle\langle\psi|$.

2. Basic Principles of Transport through a Single-Level Dot with Impurity

In other words, it is a necessary condition for the existence of an unique stationary state $\hat{\rho}_{ST}$, that the system can "forget" its initial one. If it is met, the nonequilibrium situation (described by bias voltages, temperature, etc.) completely determines the stationary physics and, hence, any reference to $\hat{\rho}_i$ should be avoided.

Nevertheless, in some cases there are good reasons to include information about the initial state into the derivation even of a stationary state $\hat{\rho}_{ST}$. The standard example is a quantum system, whose Hamiltonian $\hat{H}$ can be divided into two parts: $\hat{H} = \hat{H}_0 + \hat{H}_{INT}$, where $\hat{H}_0$ represents a known, solved problem and $\hat{H}_{INT}$ describes some (often, but not necessarily, small) deviation from it.[4] The eigenstates of $\hat{H}_0$—and only they—are available ab initio; new physics emerges by the addition of $\hat{H}_{INT}$. Growing from zero to its full value, an artificial time dependence or switching is attached to $\hat{H}_{INT}$ between initial time t_i and some evaluation time t_{EV}. In doing so, the conceptual discrimination between system parts can be advantageous in two ways. First, it provides a convenient reference point for calculations and, second, it may also foster to form intuitive explanations of the physics at hand, since the full dynamics is systematically developed in the well-understood terms of $\hat{H}_0$. The resulting resemblance to some hypothetical experimental procedure provides us with a physically consistent analogy as a foundation for the mathematical derivations. If the switching is adiabatic and the proper stationary limit performed, this procedure results in a stationary state $\hat{\rho}_{ST}$ that is still connected to the conveniently chosen initial (non-interacting) state and therefore provides the above mentioned benefits.

All this applies particularly to perturbation theory: when in our model (2.1), for example, the tunnel coupling (2.1f) is weak and correlations with leads can be neglected, the dot's stationary state may still be well-described

---

[4] If $\hat{H}_0$ is a single-particle Hamiltonian, it is usually referred to as the non-interacting or *free* part, hence, $\hat{H}_{INT}$ is called interaction part.

2. Basic Principles of Transport through a Single-Level Dot with Impurity

in terms of its non-interacting eigenstates. This can be seen in sections 2.3 and chapter 3. Yet, the procedure is not restricted to weak interaction. Nonperturbative approaches like those based on the renormalisation group, to pick just one, often maintain such a description, too—though with renormalised state quantities (for references, see chapter 4.2). Lastly, the mathematical connection between $\hat{\rho}_i$ and $\hat{\rho}_{ST}$ facilitates the discovery of transitions in parameter space between regions, in which the stationary state is determined by different physical processes. This is shown in our investigation of the single-level dot's cotunnelling regime in section 3.2.

Recapitulation: the Equilibrium Case

To understand the peculiarities of the Keldysh technique, it is helpful to compare it to the procedure of obtaining the equilibrium expectation value of an observable $\hat{O}$. The system, for which it is evaluated, shall be represented by a Hamiltonian of the form $\hat{H} = \hat{H}_0 + \hat{H}_{\text{INT}}$. In general, an expectation value of $\hat{O}$ in state $\hat{\rho}$ is given by

$$\langle \hat{O} \rangle(\hat{\rho}) = \text{Tr}\{\hat{\rho}\hat{O}\}, \tag{2.3}$$

the grand canonical equilibrium- or *thermal* state $\hat{\rho}_{\text{Eq}}(\hat{h})$ of a system with Hamiltonian $\hat{h}$ by

$$\hat{\rho}_{\text{Eq}}(\hat{h}) = \hat{X}/\text{Tr}\,\hat{X} \quad \text{with} \quad \hat{X} = \exp\{-\beta(\hat{h} - \mu\hat{N})\}, \tag{2.4}$$

where $\hat{N}$ is the particle number operator. Every such thermal state $\hat{\rho}_{\text{Eq}}(\hat{h})$ is, by definition, stationary and unique. A switching of $\hat{H}_{\text{INT}}$ that connects the states $\hat{\rho}(t_i) \equiv \hat{\rho}_{\text{Eq}}(\hat{H}_0)$ and $\hat{\rho}(t_{\text{EV}}) \equiv \hat{\rho}_{\text{Eq}}(\hat{H})$ can therefore be chosen as

2. Basic Principles of Transport through a Single-Level Dot with Impurity

adiabatic, i.e., "infinitely slow."[5] This is advantageous, because the resulting time dependence of $\hat{H}(t)$ is then purely artificial and its particular form irrelevant. Without restricting the generality of the following arguments, the initial time can be chosen to lie in the infinite past and the evaluation time can be set to zero. The dynamics of $\hat{\rho}(t)$ is given by the von Neumann equation $\partial_t \hat{\rho}(t) = -i/\hbar[\hat{H}(t), \hat{\rho}(t)]$ (see [160, 163], for example). Formal integration yields the solution $\hat{\rho}(0) = \hat{U}(0, -\infty)\hat{\rho}(-\infty)[\hat{U}(0, -\infty)]^\dagger$ with the *time evolution operator*

$$\hat{U}(t, t') = \hat{\mathbb{1}} - \frac{i}{\hbar} \int_{t'}^{t} dt_1\, \hat{H}(t_1) + \left(\frac{-i}{\hbar}\right)^2 \int_{t'}^{t} dt_1 \int_{t'}^{t_1} dt_2\, \hat{H}(t_1)\hat{H}(t_2) + \ldots$$
$$=: \hat{T} \exp\left(-\frac{i}{\hbar} \int_{t'}^{t} \hat{H}(\tilde{t})d\tilde{t}\right).$$
(2.5)

From its unitarity and the property $\hat{U}(t, t) \equiv \hat{\mathbb{1}}$ follows $\hat{U}(t', t) = [\hat{U}(t, t')]^\dagger$. With the second equality in (2.5), a notation is defined by introducing the Dyson time ordering operator $\hat{T}$, which arranges all operators of a product in ascending order from right to left. Details can be seen in Appendix A. The equilibrium (subscript 'Eq') expectation value of $\hat{O}$ with full interactions $\hat{H}_{\text{INT}}$ can then formally be written without any reference to $\hat{\rho}(0)$:

$$\langle \hat{O} \rangle_{\text{Eq}}^{\text{INT}} = \text{Tr}\{\hat{U}(-\infty, 0)\, \hat{O}\, \hat{U}(0, -\infty)\hat{\rho}(-\infty)\}, \qquad (2.6)$$

where we rearranged the operator product by using the invariance of the trace with respect to cyclic permutations.

The r.h.s. of equation (2.6) describes the following procedure: propagate every vector $|\psi\rangle$ of a suitable Hilbert space basis forward from time $-\infty$

---

[5] In fact, it is the defining property of an adiabatic process, that it is slow enough to transport an equilibrium state along the corresponding path in parameter space.

2. Basic Principles of Transport through a Single-Level Dot with Impurity

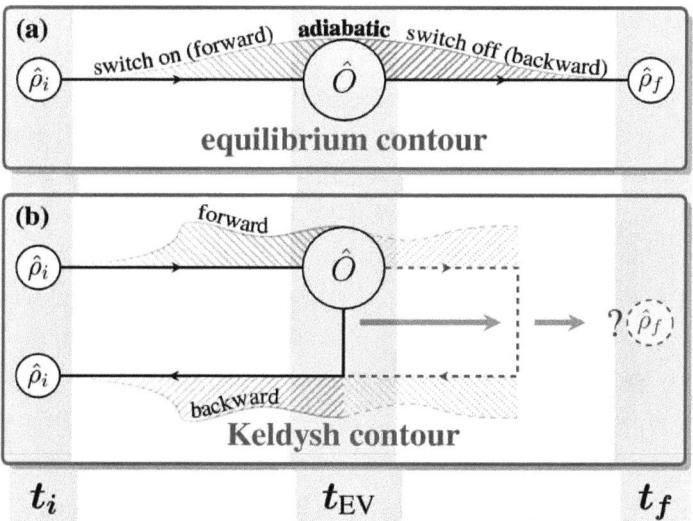

Figure 2.3.: Sketch of the time contours used to calculate the expectation value of observable $\hat{O}$. **(a)** The equilibrium contour starts at initial time $t_i(=-\infty)$ with the system being in a non-interacting initial state $\hat{\rho}_i$. Between t_i and evaluation time t_{EV}, the interaction is switched on *adiabatically* (indicated by line and dashing), so that the system is *at all times* in a (time dependent) equilibrium state. After the state with full interaction is reached at t_{EV}, operator $\hat{O}$ is applied. The *backward switch-on* from equation (2.6) is replaced by a *forward switch-off* that propagates the interacting- into the final state $\hat{\rho}_f$. The procedure rests upon the assumption, that final and initial Hilbert space vectors are physically identical. **(b)** In case of a general switching and genuine time dependence of the Hamiltonian, the *Keldysh contour* has to be used to calculate $\langle\hat{O}\rangle(t_{\text{EV}})$. Its shape follows directly from equation (2.8). Forward and backward propagation are inverse to each other; Hilbert vectors $|\psi\rangle$ acquire no phase factor. Further insertion of $\hat{\mathbb{1}} \equiv \hat{U}(t_{\text{EV}}, t_f)\hat{U}(t_f, t_{\text{EV}})$ extends the contour to some later time t_f (dotted dashed regions). The only way to obtain the corresponding state $\hat{\rho}_f$ is *a posteriori* (indicated by the question mark) and requires to solve the nonequilibrium problem after fixing $\hat{\rho}_i$.

2. Basic Principles of Transport through a Single-Level Dot with Impurity

to 0, apply operator $\hat{O}$, and propagate backwards in time to $-\infty$ before projecting on dual the vector $\langle\psi|$. For all $|\psi\rangle$, the results are summed up and weighted according to initial state $\hat{\rho}(-\infty)$. With this, the desired connection between $\langle\hat{O}\rangle_{\text{Eq}}^{\text{INT}}$ and the initial state $\hat{\rho}(-\infty)$ is established. But there is a price to be paid in form of the double time propagation. In equilibrium, there is, however, a way to get around the backward propagation by reinterpreting the *backward switch-on* between times 0 and $-\infty$ as a *forward switch-off* from 0 to $+\infty$. Since all switchings are adiabatic, it is assumed, that the final state $\hat{\rho}(t_f)$ at $t_f = +\infty$ is (physically) identical to the initial state. In other words, the time evolution from $-\infty$ to $+\infty$ is equal to multiplication with an overall phase factor: $\hat{U}(+\infty, -\infty) = e^{i\varphi}$, where φ is real. Upon insertion of $\hat{1} = \hat{U}(-\infty, +\infty)\hat{U}(+\infty, -\infty)$ at the very left in the trace in (2.6), we obtain

$$e^{i\varphi}\langle\hat{O}\rangle_{\text{Eq}}^{\text{INT}} = \text{Tr}\{\hat{U}(+\infty, 0)\,\hat{O}\,\hat{U}(0, -\infty)\hat{\rho}(-\infty)\} \\ =: \text{Tr}\{\hat{T}\hat{U}(+\infty, -\infty)\hat{\rho}_{\text{Eq}}(\hat{H}_0)\hat{O}_{t_{\text{EV}}=0}\}, \quad (2.7)$$

where the subscript of $\hat{O}$ denotes, which time is assigned to it (for the sake of ordering by operator $\hat{T}$). The resulting equilibrium time contour is shown in figure 2.3(a). It requires knowledge of the states at both ends of the contour, with the implicit involvement of $\hat{\rho}(t_f)$ manifesting itself in the factor $e^{i\varphi}$.

The Keldysh Contour

In case of a genuine time dependence of the Hamiltonian $\hat{H}(t)$, it is not possible to bypass the backward time evolution. The reason is, that an assumption similar to the equilibrium case, regarding the form of the final state $\hat{\rho}(t_f)$, can not be made. As we stated above, determining the nonequilibrium dynamics is equivalent to solving a first order differential equation in time: *one* (initial) state has to be fixed; everything else, including the

2. Basic Principles of Transport through a Single-Level Dot with Impurity

final state, should uniquely follow from the dynamical conditions given by $\hat{H}(t)$. The final state depends on the whole history of the system and can therefore consistently be fixed (in addition to the initial one) only by actually solving the nonequilibrium problem. Hence, for the calculation of the nonequilibrium expectation value $\langle \hat{O} \rangle (t_{\mathrm{EV}})$, we are stuck with both time evolutions and an equation similar to (2.6):

$$\langle \hat{O} \rangle (t_{\mathrm{EV}}) = \mathrm{Tr}\{\hat{U}(t_i, t_{\mathrm{EV}})\, \hat{O}\, \hat{U}(t_{\mathrm{EV}}, t_i) \hat{\rho}(t_i)\}. \qquad (2.8)$$

Note, that no (not even implicit) reference to $\hat{\rho}(t_f)$ is made. We end up with a forward-backward time contour, the so-called *Keldysh contour*, which is shown in figure 2.3(b). Since the forward evolution with $\hat{U}(t_{\mathrm{EV}}, t_i)$ and corresponding backward part are inverse mathematical operations, no phase factor is acquired by any vector $|\psi\rangle$, when propagated along the whole contour.

For the sake of mathematical convenience, the contour can be extended to reach some later time t_f by inserting the unity operator in (2.8) on the left of $\hat{O}$. Assuming further, that

(2.K1) $\hat{H}(t) \equiv \hat{H}(t_i)$ for $-\infty < t \leq t_i$ **and**

(2.K2) $\hat{\rho}(t_i)$ is a mixture of eigenstates of $\hat{H}(t_i)$,

allows to trivially extend the contour to the infinite past with $\hat{\rho}(-\infty) = \hat{\rho}(t_i)$. In the limit $t_f \to +\infty$, this yields

$$\begin{aligned}\langle \hat{O} \rangle (t_{\mathrm{EV}}) &= \mathrm{Tr}\{\hat{U}(-\infty, +\infty)\hat{U}(+\infty, t_{\mathrm{EV}})\, \hat{O}\, \hat{U}(t_{\mathrm{EV}}, -\infty)\hat{\rho}(-\infty)\} \\ &=: \mathrm{Tr}\{\hat{T}_{\mathrm{K}} \hat{U}_{\mathrm{K}} \hat{\rho}(-\infty) \hat{O}_{t_{\mathrm{EV}}}\},\end{aligned} \qquad (2.9)$$

where $\hat{U}_K = \hat{U}(-\infty, +\infty)\hat{U}(+\infty, -\infty)$ is the evolution operator along the whole, infinite Keldysh contour and $\hat{T}_K$ the corresponding time ordering operator. Analogous to $\hat{T}$, it arranges operator products with ascending *Keldysh times* from right to left. The Keldysh time distance between two points on the contour equals the length of the contour segment that connects them. Its sign is positive, when moving according to the propagation direction on each respective branch [indicated by arrows in figure 2.3(b)] and negative otherwise. As a consequence, points on the backward branch are later in Keldysh time than every point on the forward branch. Note, that, in general, supposition (2.K2) implies knowledge of the eigenstates of $\hat{H}(t_i)$. It therefore acts as a counterpart to the respective assumption made for equilibrium systems, namely that $\hat{H}_0$ represents an already solved problem.

There is an ambiguity in our subscript notation, which gives the position of $\hat{O}_{t_{\text{EV}}}$, as "seen" by the time ordering operator in equation (2.9). It denotes a *real-time* and, as our derivation went, $\hat{O}$ came to "sit" on the forward branch. But there are two contour positions assigned to every real-time position—one on each branch—, while $\hat{T}_K$ only sees Keldysh times. Throughout this thesis, we resolve this ambiguous situation by fixing every observable on the forward branch by default. From the physical point of view, however, the choice of the contour branch is irrelevant.

The Keldysh Partition Function

At the beginning of this section, we introduced the equilibrium partition function Z and mentioned its role as a generating function for expectation values of observables using the example of the total particle number $N \equiv \langle \hat{N} \rangle$. By comparing the identities (2.4), (2.3), and (2.2), the grand canonical partition function can be identified with the normalisation constant of the

2. Basic Principles of Transport through a Single-Level Dot with Impurity

thermal state $\hat{\rho}_{\text{Eq}}(\hat{h})$:

$$Z = \text{Tr}(\exp\{-\beta(\hat{h} - \mu\hat{N})\}). \tag{2.10}$$

But using of Z similar as in equation (2.2) is not restricted to quantities such as particle number or thermodynamic energy. By adding the source term $-\eta\hat{O}$ to the Hamiltonian in (2.10), we can take advantage of its exponential form, to transform the partition function into a generating function $Z[\eta]$ for an arbitrary expectation value $\langle\hat{O}\rangle_{\text{Eq}}$. The latter, we can then write as $\langle\hat{O}\rangle_{\text{Eq}} = \text{Tr}\{\hat{\rho}_{\text{Eq}}(\hat{h})\hat{O}\} \equiv \partial_\eta \ln Z[\eta]/\beta|_{\eta=0}$. Obviously, the partition function is obtained from $Z[\eta]$ for $\eta = 0$.

With the remarks presented in the previous paragraphs, we are now able to motivate the definitions of corresponding partition- and generating functions $\mathcal{Z}[\eta]$ for the nonequilibrium case. The starting point is equation (2.9). We compare it with the respective equilibrium equation, which we get by inserting (2.4) into (2.3):

$$\langle\hat{O}\rangle_{\text{Eq}} = \underbrace{\frac{\text{Tr}\{e^{-\beta(\hat{h}-\mu\hat{N})}\hat{O}\}}{Z[\eta=0]}}_{(2.\text{Z2})} \overset{(2.\text{Z1})}{\underset{(2.\text{Z3})}{}} \frac{\text{Tr}\{\hat{T}_K \hat{U}_K \hat{\rho}(-\infty)\hat{O}_{t_{\text{EV}}}\}}{1} = \langle\hat{O}\rangle(t_{\text{EV}}).$$

This provides us with the following analogies for the construction of $\mathcal{Z}$.

(2.Z1) The role of the Boltzmann exponential factor in equilibrium should be taken by Keldysh propagator $\hat{U}_K$ in the expression for $\langle\hat{O}\rangle(t_{\text{EV}})$. A suitable source term $\sim \eta$ (see below), added to $\hat{h}$, will again make $\hat{O}$ appear in the right place in the expectation value after differentiation with respect to η.

2. Basic Principles of Transport through a Single-Level Dot with Impurity

(2.Z2) It is usual and convenient, to normalise the Keldysh generating function $\mathcal{Z}[\eta]$ in such a way, that $\mathcal{Z}[\eta = 0] = 1$.

(2.Z3) Though present in the respective expressions for the equilibrium and Keldysh expectation values, an observable $\hat{O}$ other than $\hat{h}$ should not appear in the partition function $\mathcal{Z}[0]$ itself.

Points (2.Z1) and (2.Z3) suggest, that the Keldysh partition function should be proportional to $\text{Tr}\{\hat{U}_K \hat{\rho}(-\infty)\}$.[6] From point (2.Z2) and the observations $\hat{U}_K \equiv \hat{\mathbb{1}}$ and $\text{Tr}\{\hat{\rho}(-\infty)\} = 1$, we see, that no additional normalisation factor is needed. In view of these arguments, the definition of Kamenev and Levchenko [86] of the Keldysh partition function

$$\mathcal{Z}[0] := \text{Tr}\{\hat{U}_K \hat{\rho}(-\infty)\} = 1 \qquad (2.11)$$

provides a consistent and convenient starting point for the nonequilibrium approach we employ in chapter 4. Also, the seemingly trivial form of $\mathcal{Z}$ should not lead to irritations. $\hat{U}_K$ is only identical to $\hat{\mathbb{1}}$, when no source term is present ($\eta = 0$) and both forward and backward time evolutions are symmetrical. For non-vanishing η, this symmetry is broken and $\mathcal{Z}[\eta]$ will generally differ from 1. Still, due to the presence of the exponential evolution operator, expression (2.11) proves to be useful even in itself. A path integral representation of this partition function will allow for straightforward addition of source terms and derivations with respect to η. This is shown in section 4.1.

Let us close this section with a few words on how the source terms are introduced. With the Keldysh contour $\mathcal{K}$, evolution operator $\hat{U}_K$ can be written as

---

[6] In this expression, the operators are in correct Keldysh time order and $\hat{T}_K$ can be omitted.

2. Basic Principles of Transport through a Single-Level Dot with Impurity

$$\hat{U}_K = \hat{U}(-\infty, +\infty)\hat{U}(+\infty, -\infty) = \hat{T}_K \exp\left(-\frac{i}{\hbar}\int_K \hat{h}(t)\,\mathrm{d}t\right). \quad (2.12)$$

Kamenev and Levchenko suggest to add a term $i\hbar\hat{O}\eta(t)$ to the Hamiltonian, so that the sought-after expectation value is obtained from the resulting generating function $\mathcal{Z}[\eta(t)]$ by means of the *functional derivative*

$$\langle\hat{O}\rangle(t_{\text{EV}}) = \left.\frac{\delta \ln \mathcal{Z}[\eta(t)]}{\delta\eta(t_{\text{EV}})}\right|_{\eta=0}. \quad (2.13)$$

Note, that the logarithm is kept for the sake of a stronger resemblance to the equilibrium case (see above). Due to the normalisation property (2.Z2), it can just as well be omitted. We will show in section 4.1.7, however, that the addition of source terms is a straightforward procedure, once a path integral representation of $\mathcal{Z}$ is derived. The functional derivative in (2.13) can then be reduced to an ordinary derivative with respect to a real number η.

2.3. Preliminary Remarks to Sequential and Cotunnelling Transport

During the previous introductions of the model system and the concepts that form the foundation of our theoretical approaches, though mentioned abundantly, the notion of quantum transport itself remained abstract and in-explicit. Aim of this section is to exemplary illustrate transport in the regime of weak coupling and small temperatures. Since it is dominated by incoherent, sequential tunnelling processes, finding intuitive explanations for elementary transport effects in this regime is comparably easy. Along the way, we introduce the vocabulary of the transport language that will be used widely in the remaining parts of this work.

2. Basic Principles of Transport through a Single-Level Dot with Impurity

The process when an electron that tunnels from a lead onto the dot or back is accompanied by a charge and spin transfer. Hence, single-electron tunnelling processes, as defined by the coupling Hamiltonian (2.1f), are the elementary building blocks for a theory of charge, spin, and energy transport.[7] But although a transport phenomenon is ultimately formed by an interplay of elementary and complex charge-moving tunnelling events, for the purposes of this thesis, we do not consider an individual event as transport all by itself. Rather—as we use the concept here—transport is a dynamical process, in the course of which charge, energy, and/or spin are transferred *on average* in a quantumstatistical sense. According to this convention, individual tunnelling events appear on the level of small fluctuations; "transport", we will call only the non-vanishing, combined statistical dynamics of a macroscopic number of them. In doing so, we ensure that "current," in the sense we use the notion here, is the effect of a nonequilibrium situation in a macroscopic system.

2.3.1. Transition Rates and Rate Equations

A very intuitive and common way of describing the macroscopic, average dynamics of a system is based on *transition rates*. Each rate is an aggregated quantity giving the average number of transitions from matrix elements per time $\rho_\phi^\psi(t) := \langle\phi|\hat{\rho}(t)|\psi\rangle$ to $\rho_{\phi'}^{\psi'}(t+\mathrm{d}t)$, where $|\phi\rangle$ and $|\psi\rangle$ are Hilbert vectors of a suitable basis and $\hat{\rho}(t)$ is the time dependent density matrix. Aim of this approach is to map the system dynamics on a *continuous-time Markov process* (see, e.g., [160, 162, 164]) which is memoryless in time. That is to say, for a Markov process, it is sufficient to know the system's state at one time t, to know it in the future $t' > t$. If such a mapping is

---

[7]At least for systems with a similar structure as the one shown in figure 2.1a that can be described as an open quantum system with a coupling like $\hat{H}_T$.

2. Basic Principles of Transport through a Single-Level Dot with Impurity

possible, rates are obtained, which are immediately connected to the time derivative of the system state $\hat{\rho}(t)$ and determined by the combined statistical dynamics of all *irreducible* processes that cause the corresponding transitions. In case of our SLQD model, the transitions are caused by the electron tunnelling. A complex process is called reducible if it can be decomposed into several independent events connected by periods of (tunnelling-) free propagation. The rates of all possible transitions define a fourth rank tensor $\hat{W}(t)$, that leads to the *rate or master equation* [46, 162, 164]

$$\overbrace{\frac{\mathrm{d}\hat{\rho}}{\mathrm{d}t}(t) = \hat{W}(t)\hat{\rho}(t)}^{\text{operator notation}} \qquad \overbrace{\forall \psi, \phi : \frac{\mathrm{d}}{\mathrm{d}t}\rho_\phi^\psi(t) = \sum_{\phi',\psi'} W_{\phi \leftarrow \phi'}^{\psi \leftarrow \psi'}(t) \rho_{\phi'}^{\psi'}(t),}^{\text{tensor notation}} \quad (2.14)$$

which allows to determine the dynamical state $\hat{\rho}(t)$ and, based on that, the transport behaviour. The whole approach rests on a few assumptions besides the Markov property [162]. For this introduction, the time scale, on which the system state changes, is assumed to be much larger than the one for individual transitions.[8] Also, the transition rates should only depend on and inherit their time dependence from the Hamiltonian of the system and not its dynamical state. In the regime of weak tunnel coupling and low temperature ($|\gamma|, k_\mathrm{B}T \ll \Delta, U$), these conditions are often met to an extent that a reasonable up to excellent agreement of theory and experimental results can be reached.

In section 2.2.1, we introduced the open quantum system and the equilibrium approximation for the leads. The latter manifests mathematically in the factorisation of the density matrix $\hat{\rho}(t)$ into time independent, equilibrium lead states $\hat{\rho}_p := \hat{\rho}_\mathrm{Eq}(\hat{H}_p)$ [see equation(2.4) and (2.1e)] and the so-called *reduced density matrix* $\hat{P}(t)$ that contains the complete dynam-

---

[8]This is yet another common infinity limit idealisation.

2. Basic Principles of Transport through a Single-Level Dot with Impurity

ics of the system: $\hat{\rho}(t) = \hat{P}(t)\hat{\rho}_{\text{leads}}$ with $\hat{\rho}_{\text{leads}} := \hat{\rho}_L\hat{\rho}_R$. In general, the reduced density matrix, which is obtained by tracing out the lead degrees of freedom, has the form

$$\hat{P}(t) \equiv \text{Tr}_{\text{leads}}\, \hat{\rho}(t) = \sum_{\chi,\chi'} P^\chi_{\chi'}(t)|\chi'\rangle\langle\chi|, \qquad (2.15)$$

where $\chi \in \{0, \uparrow, \downarrow, d\}$ denotes one of the four dot eigenstates.

For this short introduction, however, we focus on the stationary state dynamics and the limit of sequential transport. The rates are then solely determined by incoherent tunnelling events of single electrons, which corresponds to perturbation theory of lowest (second) order in the amplitude γ. By purely sequential dynamics, an initially diagonal state $\hat{P}_i$ cannot acquire coherences and $\hat{P}(t)$ stays diagonal for $t > t_i$. We assume further that an unspecified source of dephasing would destroy, during the system's convergence to stationarity, any coherence present in a general state $\hat{P}_i$. In these circumstances, the stationary reduced density matrix $\hat{P}^{\text{st}}$ is diagonal. Its elements P^{st}_χ can be identified with the occupation probabilities of the states $|\chi\rangle$. In the stationary limit, equation (2.14) becomes the self-consistent, reduced linear matrix equation

$$\hat{W}^{(1),\text{st}}\hat{P}^{\text{st}} = 0 \qquad \forall \chi : \sum_{\chi'} W^{(1),\text{st}}_{\chi\leftarrow\chi'} P^{\text{st}}_{\chi'} = 0. \qquad (2.16)$$

Due to the assumption of $\hat{P}^{\text{st}}$ being diagonal, it has a vector- and $\hat{W}^{(1),\text{st}}$ has a matrix structure. The superscript integer indicates that only processes with exactly one tunnelling electron are taken into account for the calculation of the rates. For the rest of this section, we will discuss the sequential, stationary regime only and omit the superscript. We explained on page 41 in section 2.2.2, that this equation is only valid if the system can "forget" the

2. Basic Principles of Transport through a Single-Level Dot with Impurity

initial state in the course of purely sequential evolution, which is consistent with the Markovian assumption.[9]

Since all information about the initial state is lost in the stationary limit, occupation vector $\hat{P}$ is obtained by solving the system of linear equations (2.16) and, hence, determined solely by the dynamical conditions that are given by the Hamiltonian and are encoded into the transition rates. It implicitly characterises a proper stationary vector $\hat{P}$ as being unchanged by the interplay of all (statistical) tunnelling events during an infinitesimal time interval dt. This is the meaning of the attribute "self-consistent" in this context.

The overall probability of finding the dot in one of its four eigenstates is a conserved quantity and is equal to 1, while a product $W_{\chi \leftarrow \chi'} P_{\chi'}$ with $\chi \neq \chi'$ can be interpreted as the amount of probability flowing out of $|\chi'\rangle$ into $|\chi\rangle$. Hence, to keep the balance, the sum of a diagonal transition matrix element $W_{\chi' \leftarrow \chi'}$ and the rates $W_{\chi \leftarrow \chi'}$ for transitions from $|\chi'\rangle$ into *other* states $|\chi\rangle$ has to vanish:

$$\forall \chi' : \sum_{\chi} W_{\chi \leftarrow \chi'} = 0. \qquad (2.17)$$

In other words, the values of the diagonal elements are constrained by the conservation of probability. As a result, the rank of $\hat{W}$ cannot exceed $n-1$ with $n = 4$ being the dimension of the dot's Hilbert space, which allows for the existence of non-trivial solutions of the homogeneous rate equation. What is not encoded into equation (2.16), however, and was lost in the stationary limit, is the proper normalisation of the occupation vector. We have to add the corresponding condition

---

[9] In our detailed discussion of the cotunnelling regime in chapter 3, we present an example, for which this particular requirement is not fulfilled.

2. Basic Principles of Transport through a Single-Level Dot with Impurity

$$\sum_\chi P_\chi = 1 \qquad (2.18)$$

separately, to obtain a set of equations that determines $\hat{P}$ uniquely.

Charge Current

In this intuitive picture of occupation flowing between states due to single-electron tunnelling processes, a rate equation for the charge current I can be constructed in a simple manner. Throughout this thesis, we define a current of positive charges flowing from lead L to R as positive. At the same time, it is the product of the negative electron charge e with the *particle current*. If considered as particles, *electrons* tunnelling, e.g., from L onto the dot are therefore counted *positive*. Consequently, each tunnelling process that contributes to a transition rate $W_{\chi \leftarrow \chi'}$ comes with a sign, that depends on the spatial direction of the tunnelling; for a particle that tunnels through barrier p it equals to p times $S(\chi, \chi') := \text{sign}[N(\chi) - N(\chi')]$, where $N(\chi)$ is the number of particles in state $|\chi\rangle$. The charge current I is then given by

$$I = e\,\text{Tr}\{\hat{W}^I \hat{P}\} \qquad I = e \sum_{\chi\chi'} W^I_{\chi \leftarrow \chi'} P_{\chi'} \equiv e \sum_{\chi\chi'} (W^I)^{\chi \leftarrow \chi'}_{\chi \leftarrow \chi'} P^{\chi'}_{\chi'} \qquad (2.19a)$$

with

$$W^I_{\chi \leftarrow \chi'} := \frac{S(\chi, \chi')}{2}(W^L_{\chi \leftarrow \chi'} - W^R_{\chi \leftarrow \chi'}) \quad \text{and} \quad W_{\chi \leftarrow \chi'} = \sum_p W^p_{\chi \leftarrow \chi'}. \qquad (2.19b)$$

The rates $W^p_{\chi \leftarrow \chi'}$ only include processes of particles that tunnel through barrier p. In the operator version of (2.19a), as in any other similar context,

2. Basic Principles of Transport through a Single-Level Dot with Impurity

$\hat{P}$ and $\hat{W}^I$ should be interpreted as diagonal density matrix and fourth rank tensor, rather than as vector and matrix, respectively. The factor $1/2$ in definition (2.19b) is a consequence of the charge conservation, as expressed by the *continuity equation*: since the current through barrier L equals the one through barrier R, the factor prevents a double counting of the current, while the minus sign accounts for the opposite relative directions. This corresponds to the definition $I := (I^L - I^R)/2$, where I^p is the charge current flowing from lead p *into* the dot. A slightly different but equivalent definition is used in chapter 3.

Sequential Transition Rates

In the sequential regime, the rates for transitions between dot states χ and χ' can be calculated with the help of *Fermi's golden rule* (see, [165, 166], for example). For two Hilbert states $|\psi\rangle$ and $|\psi'\rangle$ of the uncoupled system $\hat{H}_0 := \hat{H}^{el}_{dot} + \hat{H}_{leads}$, the first-order transition rate in $\hat{H}_T$ is given by

$$W_{\psi \leftarrow \psi'} = \frac{2\pi}{\hbar} |\langle \psi | \hat{H}_T | \psi' \rangle|^2 \, \delta(E_\psi - E_{\psi'}), \quad (2.20)$$

where $|\psi\rangle = |\chi\rangle \otimes |\mathbf{k}_L \sigma_L\rangle \otimes |\mathbf{k}_R \sigma_R\rangle$ and $E_\psi = \epsilon_\chi + \epsilon_{k_L} + \epsilon_{k_R}$ the corresponding total energy. The Dirac delta function δ ensures the conservation of energy. To go from this expression to a rate $W_{\chi \leftarrow \chi'}$ of transitions between two states in the isolated dot's Hilbert space (the reduced state space), we have to sum over all combinations of lead components in $|\psi\rangle$ and $|\psi'\rangle$, weighting each addend according to the thermal distribution $\hat{\rho}_{leads}$. The summation over the spin can be carried out right away, since it is conserved by $\hat{H}_T$ and uniquely determined by the final states of sequential transitions in a SLQD. In case of $\chi = \uparrow$ and $\chi' = d$, for instance, the spin of the tunnelling electron has to be $\sigma = \downarrow$. In particular, note, that for all χ, χ', the

2. Basic Principles of Transport through a Single-Level Dot with Impurity

spectral weights $\langle \chi | \hat{d}_\sigma^{(\dagger)} | \chi' \rangle$ are either 0 or 1. This is also a reason why the phase of the tunnel amplitude $\gamma_k = \gamma$ can be neglected (see section 2.1.2). How to calculate the resulting trace over the lead degrees of freedom

$$W_{\chi \leftarrow \chi'} = \frac{2\pi}{\hbar} \sum_{k_L k_R} \sum_{k'_L k'_R} \langle \psi' | \hat{H}_T | \psi \rangle \langle \psi | \hat{H}_T \hat{\rho}_{\text{leads}} | \psi' \rangle \delta(E_\psi - E_{\psi'}), \quad (2.21)$$

is explicitly shown in appendix B. To carry out the last remaining sum over the wave vectors k requires the sum to be transformed into an integral over the corresponding electron energies ϵ_k. We assume, that the density of k-states is isotropic: $\varrho(k) = \varrho(|k|)$. In any system with the number $\mathcal{N}$ of quasi-continuous states, clearly, the following identity should hold for this transformation to be well-defined:

$$\mathcal{N} = \sum_k 1 = \int_\mathcal{V} \varrho(k) dk = \frac{1}{S_d(1)} \int_{\epsilon_k^-}^{\epsilon_k^+} \varrho(\epsilon_k) d\epsilon_k \int d\Omega, \quad (2.22)$$

where $\mathcal{V}$ is the k-space volume of the system and the integration limits $\epsilon_k^\pm$ may in general depend on the space direction Ω. With the number $d\nu = S_d(|k|) \varrho(|k|) d|k|$ of states in a spherical k-space shell, the *density of states* in energy space is defined by $\varrho(\epsilon_k) = d\nu/d\epsilon_k$. Here, $S_d(r)$ with $d = 2, 3$ is the surface area of a d-dimensional sphere with radius r and the geometrical factor $S_d(1)$ accounts for the proper normalisation of the angular integral.[10] In chapter 2.1.2, however, we argued that when the density of states can be assumed constant over the energy range of electrons contributing to transport ($\varrho(\epsilon_k) \approx \varrho(\epsilon_F) = \text{const.}$), we can set $\epsilon_k^\pm$ to $\pm\infty$. This is the so-called *wide-band limit*. ϵ_F is the leads' Fermi energy for $V_{\text{bias}} = 0$. Since the tunnelling amplitude is assumed to be constant, too, the energy

---

[10]For $d = 2$ and $d = 3$, we have to insert $S_d(r) = 2\pi r$ and $S_d(r) = 4\pi r^2$, respectively.

2. Basic Principles of Transport through a Single-Level Dot with Impurity

integral in (2.21) will not depend on Ω and with $\sum_k \to \varrho(\epsilon_F) \int d\epsilon_k$, we arrive at

$$W_{\chi \leftarrow \chi'} = \frac{\Gamma}{\hbar} |\langle \chi | \hat{d}_\sigma^\dagger | \chi' \rangle|^2 \sum_p f_p^+ (\epsilon_\chi - \epsilon_{\chi'}) \\ + \frac{\Gamma}{\hbar} |\langle \chi | \hat{d}_\sigma | \chi' \rangle|^2 \sum_p f_p^- (\epsilon_{\chi'} - \epsilon_\chi), \qquad (2.23)$$

where we defined $f_p^+(\epsilon) := f(\epsilon - \mu_p)$ and $f_p^-(\epsilon) := 1 - f(\epsilon - \mu_p)$. The Fermi function $f(\epsilon) = [1 + \exp(\beta \epsilon)]^{-1}$ gives the thermal occupation for a free gas of fermions. The scalar parameter $\Gamma := 2\pi |\gamma|^2 \varrho(\epsilon_F)$ is an energy that characterises the tunnel coupling more completely than the tunnel amplitude γ, for it also includes information about the leads behind the barrier via the density of states. From now on, Γ is regarded as strength of the tunnelling interaction.

The first (second) term on the r.h.s. of equation (2.23) describes tunnelling of an electron out of (into) the leads and onto (out of) the dot. Depending on the spectral weight $|\langle \chi | \hat{d}_\sigma^{(\dagger)} | \chi' \rangle|^2$, not more than one of these terms can differ from zero; in addition to σ, the tunnelling direction of the electron is also determined by the choice of the initial and final dot states. For example, if $\chi' = 0$ and $\chi = \downarrow$, the electron has to tunnel onto the dot and the second term in equation (2.23) vanishes. The rate for tunnelling onto the dot is proportional to the occupation $f_p^+(\epsilon)$ in the leads at the energy $\epsilon = \epsilon_\chi - \epsilon_{\chi'}$ that is transferred to the dot in the process. Consequently, tunnelling into a lead requires empty states at the energy $\epsilon = \epsilon_{\chi'} - \epsilon_\chi$ transferred from dot to lead. Their number is given by $f_p^-(\epsilon)$.

In general, the differences $\mu(\chi', \chi) := \epsilon_{\chi'} - \epsilon_\chi$, where $N(\chi') = N(\chi) + 1$, between end state energies of possible sequential transitions play an important role for transport considerations and are called *transport channels*. In

2. Basic Principles of Transport through a Single-Level Dot with Impurity

our SLQD model, there are four such channels: $\mu(\tilde{\sigma}, 0) = \Phi_D + \tilde{\sigma}\Delta/2$ and $\mu(d, \tilde{\sigma}) = \mu(-\tilde{\sigma}, 0) + U$, where $\tilde{\sigma} \in \{g, e\} = \{\uparrow, \downarrow\}$ is an index, that orders the single-electron states according to their energy. In this spirit, g (e) stands for the ground (excited) state. Whenever they appear as numbers, we put g := -1 and e := 1. This definition is well-defined only when $\Delta \neq 0$. It will be used in cases, in which only relative and not absolute spin orientations are important. For example, if the degeneracy of the single-particle states is broken by an external magnetic field, the absolute spin directions of ground and excited states are interchanged by an inversion of the field. If the physics does not change by such an operation—that is to say, if it is invariant under inversion of magnetic field and all spin directions—we do not specify a particular mapping of $\{\uparrow, \downarrow\}$ to $\{g, e\}$ but use only the energy-related symbols. This is the case here and in chapter 3.

Figure 2.4 shows the so-called *transport scheme* of the SLQD model (2.1). The horizontal direction depicts a spatial dimension and is interpreted as a schematic cut through the system with left and right leads (the shaded areas L and R) on both sides of the dark grey vertical tunnelling barriers, which in turn enframe the SLQD as central region. The chemical potentials of electrons in the respective areas are given in the vertical (energy) dimension. With the exception of the tunnel coupling Γ, which is only sketched graphically as the width of the barriers, energy distances are indicated by vertical braces. The shaded areas in L and R represent the occupied states, as given by the Fermi function f at temperature T and with (electro-)chemical potentials μ_L and μ_R, respectively. In a light shaded area, all states are occupied, while occupation decreases to (very nearly) zero in a dark area, whose width scales with T. By convention, the black, dashed, zero energy lies halfway between both Fermi levels, their energy difference corresponding to eV_{bias}. The four discrete transport channels of the dot are symbolised by vertical lines in the central region.

2. Basic Principles of Transport through a Single-Level Dot with Impurity

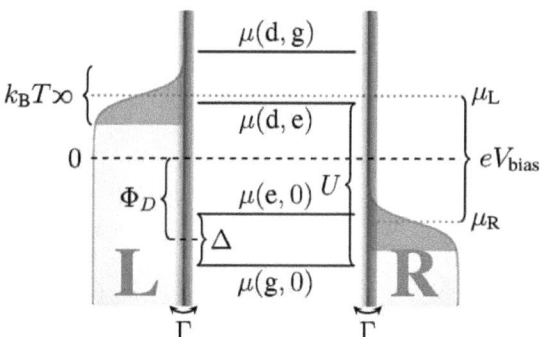

Figure 2.4.: Introduction of the transport scheme for our model (2.1) (without magnetic impurity, $\hat{H}_{\text{dot}} \equiv \hat{H}_{\text{dot}}^{\text{el}}$). The two thick vertical lines symbolise the tunnelling barriers, which separate the quantum dot as central region from the thermal leads L and R. In the central dot part, we symbolise transport channels $\mu(\chi', \chi)$ by horizontal grey lines. The label g (e) indicates the single-electron state with lower (higher) energy, as referring to the single-particle ground (excited) state. The shaded areas in L and R represent the occupied states (given by the Fermi function f, see text) of the free Fermi gases at temperature T and with (electro-)chemical potentials—also called Fermi levels—μ_L and μ_R, respectively. In the light shaded areas, all states are occupied, while occupation decreases to zero in the dark areas, whose widths scale with T. By convention, the black, dashed zero energy lies halfway between both Fermi levels, their energy difference corresponding to eV_{bias}. The tunnel coupling Γ is sketched graphically as the width of the barriers.

2.3.2. Sequential Transport and Coulomb Blockade

In the sequential transport regime, both the coupling Γ and the thermal energy $k_B T$ are much smaller than the dot energies Δ and U. As long as they are small enough to allow for a perturbative treatment as described in the previous section, occupation vector $\hat{P}$ and charge current I are obtained by plugging equation (2.23) into the rate equations (2.16) and (2.19), respec-

2. Basic Principles of Transport through a Single-Level Dot with Impurity

tively. In sequential transport, the average dynamics of the macroscopic system is caused by the interplay of incoherent one-by-one tunnelling processes only. This results in the strict energy conditions, expressed by the Fermi functions (see above).

The transport scheme in figure 2.5(a) shows a configuration, which leads to a non-zero sequential current. The lowest two transport channels lie in the energy range between the Fermi levels μ_p, called *transport window*. Since every sequential tunnelling event has to conserve the total energy, only electrons that are level with transport channels can tunnel onto the dot. In turn, tunnelling into a lead requires empty (not fully occupied) states at the level of transport channels. In the shown configuration, occupied and empty states of matching energy can only be found in lead L and R, respectively. As a result, the bias voltage imposes a preferred direction for statistical tunnelling of electrons, i.e, individual electrons are more likely (much more, in this case) to tunnel from L to R than in the reverse direction. This leads to a net charge current from L to R.

In order to contribute to a sequential current, it is therefore necessary but not sufficient for a transport channel to lie within the transport window. This is shown in figure 2.5(b) for a configuration, in which a *Coulomb blockade* inhibits sequential transport through the dot [32, 46, 61, 167]. Since channel $\mu(g, 0)$ lies below both chemical potentials μ_p and $\mu(e, 0)$ above μ_R due to $\Delta > 0$, the SLQD has to be in the single-electron Hilbert state $|g\rangle$, also referred to as (single-level) ground state. Once the system propagates into this state, which inevitably happens during its approach to stationarity, it gets trapped in it; no free states exist in the leads at the matching energies for tunnelling out. It is also said, that the lowest channel is *Pauli-blocked*. The forbidden tunnelling is indicated by dashed arrows, labelled with (i). Despite lying in the transport window, the two channels $\mu(e, 0)$ and $\mu(d, e)$ cannot contribute to transport, because tunnelling onto the dot via those

2. Basic Principles of Transport through a Single-Level Dot with Impurity

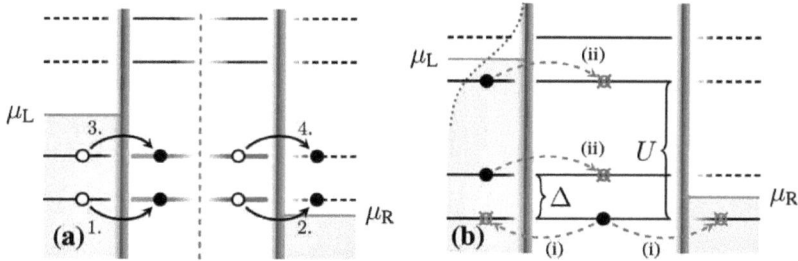

Figure 2.5.: Illustration of sequential transport and the Coulomb blockade by means of energy profiles as introduced in figure 2.4. Since the temperature is assumed to be small, the very steep Fermi functions are depicted as to be step-like around $\mu_{L/R}$. The solid (dark dashed) horizontal lines mark occupied (empty) states, out of (into) which electrons can tunnel onto (out of) the dot sequentially. Due to the conservation of energy, these lines have to be levelled with the transport channels. **(a)** If the dot is initially empty, an electron can tunnel onto the dot only from the left lead [process (1./3.)], while a dot in a single-electron state will be emptied into the right lead with unoccupied states at the same energy [process (2./4.)]. Both the holes in L and the electrons above μ_R are assumed to thermalise instantly. Hence, in the shown configuration, a preferred direction for tunnelling exists, which leads to a charge current from left to right. **(b)** Coulomb blockade of sequential transport. Though two channels lie within the transport window, no sequential current flows through the dot. Since channel $\mu(g, 0)$ lies below both Fermi levels and $\mu(e, 0)$, due to $\Delta > 0$, lies above μ_R, the dot will be trapped in state $|g\rangle$ (black dot on lowest channel). The absence of unoccupied states at matching energy prohibits tunnelling into the leads [light dashed arrows (i)]. Tunnelling onto the dot via channels $\mu(e, 0)$ or $\mu(d, e)$ requires the dot to be in one of states $|\chi = 0, e, d\rangle$, which is thus inhibited (ii). Only through the highest channel $\mu(d, g)$, tunnelling onto the dot would be possible. Due to the Coulomb energy U, however, this channel lies above the μ_p and due to the small temperature T, no states at that energy are occupied in the leads. For larger T and broader Fermi level (dotted line and light shading), the blockade is lifted.

2. Basic Principles of Transport through a Single-Level Dot with Impurity

would require the SLQD to be in one of the states $|\chi = 0, e\rangle$ [arrows (ii) in the figure].

The remaining, highest channel $\mu(d, g)$ belongs to a transition, which connects to the stationary dot state $|g\rangle$. Tunnelling onto the dot would hence be possible, if occupied electron states of matching energy could be found in the leads. But since the channel is above the Fermi levels due to the Coulomb repulsion, given by U, this is not the case. For this reason, a situation with blocked sequential transport of the kind shown in 2.5(b) is called Coulomb blockade. As a result, we can identify two requirements a channel $\mu(\chi', \chi)$ has to meet, which together are sufficient for its contribution to a sequential charge current:

(2.R1) The channel hast to lie within the transport window, that is to say,
$$\min(\mu_p) \leq \mu(\chi', \chi) \leq \max(\mu_p).$$

(2.R2) The average stationary probability to find the dot in either one of the states χ and χ' has to be non-zero.

It can also be understood from the figure, why an effective Coulomb blockade requires a low temperature. If $k_B T$ was of the same order of magnitude as the inter-channel energy distances, the much slower exponential decrease of the occupation around the μ_p would both result in a considerable occupation in L at the level of channel $\mu(d, g)$ (indicated by the dotted line and light shading) and empty states in R at $\mu(g, 0)$.

2.3.3. Beyond Sequential Transport: Cotunnelling

For the sake of simplicity and clearness, we only considered sequential tunnelling for the calculation of the transition rates so far.[11] But even when the

---

[11] In this whole section, we assume that the model meets all conditions that are sufficient to adequately describe it by a rate equation.

2. Basic Principles of Transport through a Single-Level Dot with Impurity

coupling Γ is very small compared to the dot energies Δ and U, second- and higher-order contributions in Γ may still considerably affect the transport behaviour. In particular, effects like the Coulomb blockade might require to take second-order processes into account to avoid unphysical results. We elaborate on these particular circumstances in chapter 3. For now, just a short, qualitative introduction to second-order transport is provided. A more precise and detailed discussion follows in the next chapter.

As we mentioned in section 2.3.1, transition rates $W_{\chi \leftarrow \chi'}$ are $\mu(\chi', \chi)$ lead-averaged quantities, which aggregate the statistical effect of those irreducible processes that cause transitions between dot states χ and χ'. Sequential rates like (2.23) approximate the full rate by restricting the class of process taken into account to incoherent single-electron processes. *Cotunnelling* processes, where either *two* electrons coherently tunnel once or one electron tunnels *twice*—but also coherently—through a virtual intermediate state, are the simplest examples for irreducible many-particle tunnelling. The rates that correspond to this class of processes are of second-order in Γ and may also be calculated by Fermi's golden rule, as long as the Fermi levels are not in or very close to resonance with a transport channel [168].

In the Coulomb blockade regime, sequential tunnelling is strongly (exponentially) suppressed due to the small temperature ($k_B T \ll \Delta, U$). It follows that in this situation, cotunnelling, although it scales with Γ^2, is stronger than sequential tunnelling and dominates the transport behaviour. Not only is it responsible for the leading contribution of the current I in the Coulomb blockade regime, it also determines the stationary occupation of states that are simultaneously cut-off from sequential transport, as shown in chapter 3. Figure 2.6 shows two examples for cotunnelling processes that can occur in the Coulomb blockade configuration shown in figure 2.5(b).

In 2.6(a), one electron tunnels twice from L to R via virtual intermediate state $|d\rangle$. The SLQD is assumed to be in the single-electron ground state,

2. Basic Principles of Transport through a Single-Level Dot with Impurity

so that the transport channel used has to be $\mu(d, g)$. A process of this kind is called *elastic* as the energy of the dot remains unchanged. That this is possible at all relates to the fact that the second-order process has to conserve the system energy only when regarded as a whole. The elementary sub-processes of tunnelling from L into virtual state $|d\rangle$ and further from $|d\rangle$ into lead R, violate the energy conservation for the (short) time span that the tunnelling lasts. Hence, the occupied (line hatched) and empty (dotted) lead states that are available for tunnelling onto and out of the dot, respectively, do not just lie exactly at the discrete energies that are level with the transport channels, but also in broader regions around them. According to *Heisenberg's uncertainty principle* [163, 169], however, the tunnelling probability has to decrease with larger energy distance to the channel, viz., violation of energy conservation. This is expressed by the shape of the hatched areas: the smaller their width at a certain energy the lower is the quotient of tunnelling electrons. Because the elastic process as a whole has to be energy conserving, empty states in R above μ_R are not available as final states. These excluded states are defined by the dashed line.[12]

In 2.6(b), an *inelastic* cotunnelling event is shown. The coherent two-particle process consists of one electron tunnelling from L into the dot state $|e\rangle$ and the ground state electron tunnelling from the dot into lead R. By the process, one electron is effectively transferred from L to R and the dot is excited from $|g\rangle$ to $|e\rangle$ (dashed arrow). It is this effective excitation that qualifies the cotunnelling as inelastic. The energetic rules regarding the available lead states for tunnelling are analogous to those for the elastic case. Only the fact, that the energy for the dot's excitation has to be provided by the leads, results in an energy distance of Δ between the line hatched region in L and the dotted region in R. Again, the states in the dashed outlined regions cannot contribute to transport via this process due to energy conservation.

---

[12] Although not represented in the figure, occupied states in L below μ_L are excluded, too.

2. Basic Principles of Transport through a Single-Level Dot with Impurity

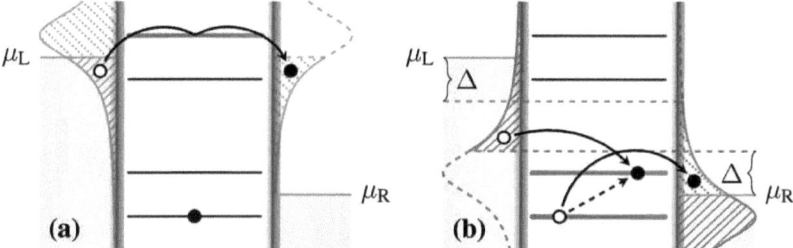

Figure 2.6.: Sketch of the two different kinds of cotunnelling processes. Since the energy configuration is the same as in figure 2.5(b), sequential transport is blocked. Cotunnelling and higher-order tunnelling, however, is possible. Areas hatched with lines (dots) represent occupied (empty) states available for transport. In contrast to the virtual, single-particle transitions that together form a coherent cotunnelling process, both these complex events (a) and (b) as a whole conserve the total energy. Hence, electrons and empty states for cotunnelling are not only available at the discrete transport channels, but also in broader regions around them. The quotient of contributing tunnelling states at a certain energy level is symbolised by the horizontal extent of the respective region. It always decreases with larger distance to the assigned channel. Dashed outlined regions are not available for cotunnelling due to the conservation of energy. **(a)** Elastic cotunnelling. By this kind of process, the energy of the dot is unchanged. The SLQD remains in the single-electron ground state, while a single electron tunnels twice: from the line hatched area in L into the corresponding dotted region in R through virtual intermediate state $|d\rangle$. **(b)** Inelastic cotunnelling. The name refers to the change in the dot energy that is caused by this process, which is indicated by the dashed arrow. Two electrons tunnel coherently and, as in (a), effectively transport one electron from L to R. Since the energy for excitation of the dot has to be provided by the leads, electrons that tunnel from L have to be at least Δ above μ_R. In turn, electrons cannot tunnel into parts of the dashed region in R that are higher than $\mu_L - \Delta$.

3

Influence of Spin Relaxation on Transport in Cotunneling Regime

The contents of this chapter have been published slightly modified in Ref. [85]: Daniel Becker and Daniela Pfannkuche, Coulomb-blocked transport through a quantum dot with spin-split level: Increase of differential conductance peaks by spin relaxation, Physical Review B 77, 205307, 2008.

COHERENCE EFFECTS, whose signatures can be seen in (spin-) electronic transport through low dimensional nanoscopic structures like quantum dots[32, 50, 79], provide insight into fundamental aspects of quantum mechanics and have important applications in vital fields of research such as spintronics, quantum computing, and data storage. [65, 68] For the occurrence of these effects spin-flip relaxation is widely considered as a limiting factor and therefore usually sought to be as small as possible. In experiments spin-flip relaxation times T_1 ranging from μs[57, 80] to ms[170, 171] have been observed displaying dependences both on the sort of quantum dot and on the parameters of the experimental setup like temperature and

3. Influence of Spin Relaxation on Transport in Cotunneling Regime

magnetic field [32, 50, 53, 58]. Accordingly, the relaxation rates can be experimentally adjusted in a wide range either by means of tuning of external parameters or by suitably tailoring the quantum dot itself. Recently, spin-flip times of even several hundred milliseconds were measured in n-doped (In, Ga)As/GaAs quantum dots charged with spin-polarized electrons at low magnetic field and temperature. [81] For transport through few-electron quantum dots in the presence of intrinsic spin relaxation, which is discussed in this chapter, microscopic mechanisms like phonon-induced spin decay due to spin-orbit or hyperfine interaction have been investigated theoretically (see, e.g., Refs. [50, 53, 58, 59, 82, 83]).

Though relaxation mostly acts destructively on coherent electron dynamics, it can also, however, considerably pronounce their effect, as Weymann and Barnaś [55] show for the case of Coulomb-blocked transport through a single-level quantum dot (SLQD) coupled to ferromagnetic leads with antiparallel magnetization. The zero-bias anomaly of the differential conductance and the conductance step at the onset of inelastic cotunneling are increased by a slow spin relaxation for a spin-degenerate and a spin-split dot level, respectively.

We will show in this chapter that a similar, strongly pronounced effect should be observable even when the leads are non-magnetic. In the considered case a small relaxation that is roughly as large as the tunnel coupling maximizes peaks of the differential conductance, which mark the onset of cotunneling-mediated sequential transport. This effect is associated with sequential tunneling out of an excited single-particle state. Within a SLQD model, the dot's level has to be spin-split. For few-electron GaAs/Al$_{0.3}$Ga$_{0.7}$As quantum dots with non-degenerate orbital levels these signatures of the single-particle spectrum have been intensely studied both experimentally and theoretically by Schleser et al. [84], whereby new insight was provided into the interplay between sequential and cotunneling in the

3. Influence of Spin Relaxation on Transport in Cotunneling Regime

Coulomb blockade regime. In the present chapter we investigate this interplay in further detail. Since said signatures appear close to resonance with single-particle transitions, we have to base our calculations on the nonequilibrium Keldysh formalism rather than second-order perturbation theory. [47, 84, 88] It is shown that in order to obtain physically correct, perturbative results for the entire cotunneling regime, one has to construct non-systematic rate equations similar to those proposed in Ref. [172]. In the latter equations we identify terms that cannot belong to the second-order perturbation expansion. By omitting these terms, one ensures that for the considered system the occupation probabilities are well-defined everywhere in the Coulomb blockade regime. As in Ref. [55] we treat the effect of relaxation phenomenologically, describing the intrinsic spin-flip processes by an effective rate θ. Thus no particular mechanism has been specified.

The chapter is structured as follows. In Sec. 3.1 we introduce the model system and explain restrictions on the system parameters. The diagrammatic transport theory and the derivation of transport equations are sketched out in Sec. 3.2. Results are presented and discussed in Sec. 3.3 and followed by a summary in Sec. 3.4.

3.1. Model

We consider a model system consisting of a SLQD, which is coupled to two metallic leads (L and R) by identical tunneling barriers, so that a dc bias voltage V_{bias}, symmetrically applied between both reservoirs, causes a tunneling current through the dot. An additional capacitatively coupled gate electrode allows to adjust the electrostatic potential Φ_D in the dot by applying a gate voltage. Such a system can be represented by the Anderson-type Hamiltonian $\hat{H} = \hat{H}_D + \hat{H}_L + \hat{H}_R + \hat{H}_T$ with the quantum dot

3. Influence of Spin Relaxation on Transport in Cotunneling Regime

part $\hat{H}_D$, the Hamiltonians $\hat{H}_L$ and $\hat{H}_R$ of the left and right lead, respectively, and the tunneling operator $\hat{H}_T$, describing the coupling between the dot and the leads. We assume that the spin degeneracy of the two single-electron dot states is lifted (e.g., by a Zeeman-field), leading to an energy difference of Δ. Then the dot Hamiltonian can be written as $\hat{H}_D = (\epsilon + \Phi_D)\hat{d}_g^\dagger \hat{d}_g + (\epsilon + \Phi_D + \Delta)\hat{d}_e^\dagger \hat{d}_e + U \hat{d}_e^\dagger \hat{d}_e \hat{d}_g^\dagger \hat{d}_g$. Here the index g (e) denotes the spin of the single-electron ground state $|g\rangle$ (excited state $|e\rangle$) and $\hat{d}_\sigma^\dagger$ ($\hat{d}_\sigma$) with $\sigma \in \{g, e\}$ creates (annihilates) an electron with spin σ and energy $\epsilon_\sigma = \epsilon + \Phi_D + \delta_{\sigma,e}\Delta$ when acting on the empty dot state $|0\rangle$ (δ is the Kronecker delta). U is the Coulomb-energy of the doubly occupied state $|d\rangle$. The leads play the role of macroscopic reservoirs and are described as free electron gases with Hamiltonian $\hat{H}_r = \sum_{k,\sigma} \epsilon_{k,r} \hat{c}_{k,\sigma,r}^\dagger \hat{c}_{k,\sigma,r}$, where $r \in \{L, R\}$ refers to the lead; k is the wave vector of an electron in reservoir r, σ is its spin and $\epsilon_{k,r}$ its energy. The $\hat{c}_{k,\sigma,r}^\dagger$ ($\hat{c}_{k,\sigma,r}$) are the corresponding creation (annihilation) operators. Due to the applied bias voltage, $\mu_r = (-1)^{\delta_{r,L}} eV_{\text{bias}}/2$—with $e > 0$ being the elementary charge— gives the electrochemical potential of reservoir r. The coupling between the leads and the dot is described by $\hat{H}_T = \sum_{k,\sigma,r}(\gamma \hat{d}_\sigma^\dagger \hat{c}_{k,\sigma,r} + \text{h.c.})$, where the first (second) term on the r.h.s. describes tunneling into (out of) the dot with the complex tunneling parameter γ (γ^*), which is assumed to be independent of wave vector and spin of a tunneling electron as well as the reservoir out of which (into which) it tunnels. With the constant density of states ρ of the reservoirs the coupling can be characterized by the positive scalar parameter $\Gamma := |\gamma|^2 \rho$ alone.[1] The stationary tunneling current I is the expectation value of the current operator $\hat{I} := \hat{I}_L$, where $\hat{I}_r = -i(e/\hbar) \sum_{k,\sigma} (\gamma \hat{d}_\sigma^\dagger \hat{c}_{k,\sigma,r} - \text{h.c.})$ [2].

---

[1] Thus, we define the coupling strength as $(2\pi)^{-1}$ times the rate for sequential tunneling (obtained by Fermi's golden rule).
[2] $I = I_L = -I_R$ due to the conservation of charge.

3. Influence of Spin Relaxation on Transport in Cotunneling Regime

We demand that the reservoirs stay in equilibrium even when coupled to the SLQD. For a perturbative calculation of the occupation probabilities and the current up to second order in the tunnel coupling, Γ has to be very small compared to the dot energies ϵ and U. Coulomb blockade of sequential transport is possible, if the thermal energy is much smaller than U, i.e., $\beta^{-1} \equiv k_B T \ll U$ with T being the temperature and k_B Boltzmann's constant. We restrict our study to parameter sets with $\beta\Gamma \ll 1$, which is a necessary condition for physical behavior of the second-order perturbation expansion in Γ once the electro-chemical potential of a reservoir is close to resonance with the energy of a single-charge excitation. [173] Furthermore, to be able to see transport signatures of the excited state within the Coulomb blockade regime, the Zeeman-splitting Δ must not be very much smaller than U but roughly of the same order of magnitude (though not larger than $U/2$). This implies $\beta\Delta \gg 1$. For the particular parameter set ($\Gamma = 4.5 \times 10^{-3} k_B T, \Delta = 45 k_B T, U = 225 k_B T$) we use throughout the following discussions, this requirement may be difficult to meet for quantum dots made of GaAs or Si and magnetic fields available in laboratories. On the other hand, the presented perturbative framework can be applied to systems with Γ that is up to 10 times larger and with U, Δ that are 10 times smaller, while yielding the same qualitative results. In practice, since it is purely of mathematical origin, the criterion $\beta\Gamma \ll 1$ is not experimentally relevant and imposes no restriction on the physics underlying the transports effects, we present here. Therefore, the results of our approach can also be applied to experiments on GaAs or Si quantum dots in which the energy difference between spin-split levels can be resolved in transport spectroscopy. [25] InAs nanowire quantum dots, however, have an effective gyromagnetic factor g between 8 and 9[34] and an effective mass of $m^* = 0.02 m_e$ (m_e is the mass of a free electron).[35] In experiments with these dots an adequately large spin-splitting should be feasible for Zeeman fields in the range of 1 to 10 T, even for the parameter set we use here.

We would also like to emphasize in this context, that the presented approach is not restricted to quantum dots with one spin-split single-particle level but can in the same way be employed for dots with two non-degenerate spinless orbitals. For very similar systems (few-electron GaAs/AlGaAs quantum dots) the discussed conductance peaks were seen in low magnetic fields. [84]

3.2. Master equations

For our calculations we use the real-time diagrammatic technique developed by Schoeller and Schön [46]. It is based on the Keldysh formalism and allows to represent dynamical, nonequilibrium properties of the model system by a formaly exact, infinite perturbation expansion with small parameter Γ. From such an expression one can obtain a systematically expanded quantity up to a finite order in the coupling.[54, 174–178] In order to construct perturbative solutions for the occupation probabilities and the tunneling current, we first assume that intrinsic relaxation is absent. To compute the time-dependent statistical expectation value $\text{Tr}\{\hat{\rho}(t)\hat{I}^r\}$ of the current operator, we have to calculate the density matrix $\hat{\rho}(t)$, which contains the complete system dynamics. Since the reservoirs are assumed to stay in equilibrium at all times, the density matrix's reservoir degrees of freedom can be integrated out using Wick's theorem. This yields the *reduced density matrix* $\hat{\rho}_D(t)$, which depends only on the dot degrees of freedom. Via an adiabatic switching between times t_0 and t the initial state of the isolated dot, represented by $\hat{\rho}_D^0 \equiv \hat{\rho}_D(t_0)$, is connected to the reduced density matrix of the coupled system $\hat{\rho}_D(t)$. This relation is expressed by equation

$$\hat{\rho}_D(t) = \hat{\Pi}(t, t_0)\hat{\rho}_D^0, \qquad (3.1)$$

3. Influence of Spin Relaxation on Transport in Cotunneling Regime

where $\hat{\Pi}(t,t')$ is a time evolution operator describing propagation of the reduced density matrix between t' and t. The propagator $\hat{\Pi}(t,t')$ can be represented as an infinite sum of diagrams on the Keldysh contour, each of which is decomposable into parts $\hat{\Pi}^0$ corresponding to propagation that is not influenced by the reservoirs and irreducible self-energy parts that describe coherent dynamics governed by the tunnel coupling and allow the dot to change its state [46]. With the operator $\hat{\Sigma}$, which consists of all irreducible diagrams, a Dyson equation for $\hat{\Pi}$ can be set up leading to the *kinetic equation*

$$\hat{\rho}_D(t) = \hat{\Pi}^0(t,t_0)\hat{\rho}_D^0 + \int_{t_0}^{t} dt_2 \int_{t_0}^{t_2} dt_1 \hat{\Pi}^0(t,t_2)\hat{\Sigma}(t_2,t_1)\hat{\rho}_D(t_1), \quad (3.2)$$

when plugged into (3.1). In the limit of $t_0 \to -\infty$ and vanishing adiabatic switching, the derivative of Eq. (3.2) with respect to t becomes a self-consistent conditional equation for the stationary reduced density matrix $\hat{\rho}_D^{st}$, provided that the SLQD will eventually forget its initial state $\hat{\rho}_D^0$ due to the interaction with the macroscopic reservoirs [3]. Since we assume diagonality of the initial density matrix $\hat{\rho}_D^0$, which implicates diagonality of $\hat{\rho}_D^{st}$ [4], it is convenient to replace the latter by the vector $\mathbf{P}$ of the stationary probabilities $P_\phi = \langle \phi | \hat{\rho}_D^{st} | \phi \rangle$ for the dot to be in state $|\phi\rangle$ with $\phi \in \{0, g, e, d\}$. We then replace the tensor operator $\hat{\Sigma}$ with the matrix $\mathbf{W}$, whose elements

$$\mathbf{W}_{\phi' \leftarrow \phi} := \int_{-\infty}^{0} dt' \Sigma_{\phi' \leftarrow \phi}^{\phi' \leftarrow \phi}(0,t') \quad (3.3)$$

are interpreted as stationary rates of quantum dot transitions from state $|\phi\rangle$ to state $|\phi'\rangle$. Since the total probability has to be conserved, the 4×4 Matrix $\mathbf{W}$ has a rank of three. Therefore the resulting self-consistent *rate equation*

$$\mathbf{WP} = 0 \quad (3.4)$$

---

[3] In the stationary state, we can then w.l.o.g. set $t = 0$.
[4] This can be seen by looking at the structure of the diagrams that constitute propagator $\hat{\Pi}$.

3. Influence of Spin Relaxation on Transport in Cotunneling Regime

has non-trivial solutions and, together with the normalization condition $\sum_\phi P_\phi = 1$, uniquely determines **P** as a vector of probabilities.[54, 172] A similar equation for the current I^r out of reservoir r into the dot can be formed, if we introduce an operator $\hat{\Sigma}^r$, which consists of all irreducible diagrams of $\hat{\Sigma}$, each having its last internal vertex replaced by an external vertex stemming from $\hat{I}^r(t=0)$. With a matrix $\boldsymbol{W}^r$, defined in analogy to (3.3), we get

$$I = -e\sum_\phi (\boldsymbol{W}^L \mathbf{P})_\phi = e\sum_\phi (\boldsymbol{W}^R \mathbf{P})_\phi. \tag{3.5}$$

Each of the Eqs. (3.4) and (3.5) yields an infinite system of coupled equations, if we express $\boldsymbol{W}$, $\boldsymbol{W}^L$, **P**, and I as expansions in Γ and sort all terms by order. The nth-order occupation vector and current are given by

$$\begin{aligned}\mathbf{0} &= \sum_{l=1}^{n} \boldsymbol{W}^{(l)} \mathbf{P}^{(n-l)} \\ I^{(n)} &= -e\sum_\phi \sum_{l=1}^{n} \left(\boldsymbol{W}^{L(l)} \mathbf{P}^{(n-l)}\right)_\phi. \end{aligned} \tag{3.6}$$

With the terms $\boldsymbol{W}^{(n)}$ and $\boldsymbol{W}^{r(n)}$ we identify those parts of $\boldsymbol{W}$ and $\boldsymbol{W}^r$, respectively, that are represented by irreducible diagrams with exactly n tunneling lines. Each of these lines connects two vertices on the Keldysh contour and represents the Wick contraction of the corresponding reservoir operators. The ascending orders of **P** and I are then calculated iteratively, starting with $\mathbf{P}^{(0)}$ and $I^{(1)}$, where the $\mathbf{P}^{(n)}$ are normalized according to $\sum_\phi P_\phi^{(n)} = \delta_{n,0}$. The first- and second-order equations describe transport caused by sequential tunneling and cotunneling processes, respectively. The curve of the sequential current against the bias voltage resembles a staircase with thermally broadened steps formed at bias values appropriate for

3. Influence of Spin Relaxation on Transport in Cotunneling Regime

single-charge excitations. Cotunneling further broadens these steps [173] and dominates the transport behavior within the *Coulomb blockade regime* (or cotunneling regime), where the gate voltage is tuned to charge the SLQD with one electron, while the bias is too small to doubly occupy or to empty the dot. [47, 168, 179, 180]

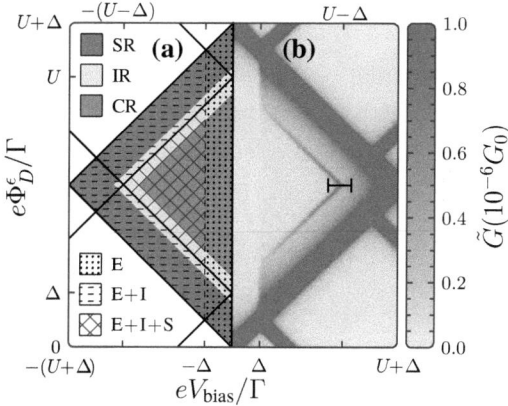

Figure 3.1.: (a) Schematic picture of the diamond-shaped cotunneling regime showing its subdivision into areas with different possible tunneling processes (hatched areas) as well as into core (CR), shell (SR), and intermediate region (IR), in which the occupations P_g and P_e are determined either by cotunneling, sequential tunneling, or both [colored (shaded) areas]. E stands for elastic and I for inelastic cotunneling, S is for sequential tunneling. (b) Calculated charging diagram [based on Eqs. (3.9) and (3.11)] of the cotunneling regime with parameters $\Delta = 45 k_B T$, $U = 225 k_B T$, $\Gamma = 4.5 \times 10^{-3} k_B T$ and $\theta = \Gamma/2$. High values of the differential conductance $\tilde{G} = d\tilde{I}/dV_{\text{bias}}$ (in units of $G_0 = \beta \Gamma e^2/\hbar$) outside and at the border of the diamond are clipped by the color scale (values near the border and in the exterior: see text). The short horizontal line corresponds to the range of bias values in Fig. 3.2. Both (a) and (b) can be extended to regions with opposite sign of eV_{bias} by reflection with respect to the $eV_{\text{bias}} = 0$ axis.

3. Influence of Spin Relaxation on Transport in Cotunneling Regime

Depending on its strength Γ, the coupling of a microscopic system like a quantum dot to macroscopic reservoirs will modify the dot's behavior slightly ($\Gamma \ll \epsilon, U$) or drastically ($\Gamma \gg \epsilon, U$). Even in the first case [62] and for incoherent sequential transport [181], the coupling may be too strong to assume that the dot propagates as if isolated between two tunneling events. In such cases and all the more when transport of highly correlated electrons is considered [182, 183], it may still be possible to represent the dot as isolated but with *renormalized* instead of the *bare* system parameters. Strictly speaking, this is also true for the system studied here and correctly accounted for by the diagrammatic technique. [62] However, since the difference between renormalized and bare energies scales with Γ, for the regime we investigate ($\Gamma \ll \beta^{-1} \ll \epsilon, \Delta, U$) it is so small, that the effect of the renormalization on the discussed transport phenomena is virtually unobservable. This can be seen, for example, in Fig. 3.1(b), where a light vertical shade within the Coulomb blockade regime indicates the onset of inelastic cotunneling as soon as $|eV_{\text{bias}}|$ equals the *renormalized* excitation energy (see below), while on the eV_{bias} axis the position of the *bare* spin-splitting Δ is marked. Obviously, both positions do not considerably deviate from each other. Hence, throughout the following discussions, we do not distinguish between the bare and renormalized quantities, although all statements are strictly valid only for the latter ones.

Fig. 3.1(a) schematically shows the (sub-)structure of the cotunneling regime plotted against eV_{bias} and the dot potential energy. For convenience the latter is given relative to the energy that is needed to charge the dot with one electron in the ground state: $e\Phi_D^\epsilon := e\Phi_D - \epsilon$. On the one hand the diamond-shaped cotunneling regime (Coulomb diamond) breaks up into three regions differing with respect to the kind of tunneling processes that predominantly determine the occupation of the single-particle states. As we discuss below P_g and P_e are given by sequential tunneling in the red

3. Influence of Spin Relaxation on Transport in Cotunneling Regime

colored (medium gray) *shell region* (SR), by cotunneling in the green (dark gray) *core region* (CR) and by a mixture of both in the yellow (light gray) *intermediate region* (IR).

On the other hand one can distinguish three sub-regimes of the Coulomb diamond with different current-driving tunneling processes. The quantum dot is in state $|g\rangle$ ($P_g \approx 1$) for $|eV_{\text{bias}}| < \Delta$ (dotted area) and the small finite current is maintained just by energy-conserving *elastic* cotunneling (E) through virtual states $|0\rangle$ and $|d\rangle$ (elastic regime). Once $|eV_{\text{bias}}|$ exceeds Δ, *inelastic* cotunneling processes can excite the dot into state $|e\rangle$ while transferring energy from the reservoirs into the SLQD [84, 167]. Since each of these processes effectively carries one electron through the dot, they cause an additional electron flow (+I). When passing from the core to the shell part of this inelastic cotunneling regime—the corresponding areas are hatched with crossed lines and dashed horizontal lines, respectively—cotunneling-mediated *sequential* tunneling out of the excited state sets on and further increases the current (+S) [84].

Before we can study the transport in vicinity of the excited state resonances, it is necessary to modify the rate equations (3.6), as they prove to be unsuitable to describe the occupations and current in the intermediate region. As Weymann et al. show in Ref. [172] it is due to the breakup of the cotunneling regime into core, shell, and intermediate region that no systematic second-order expansion of $\mathbf{P}$ or I exists, which is valid within the entire regime. This can be explained as follows.

In the shell region only those sequential transitions are energetically forbidden—the corresponding rates being exponentially small—that carry the dot out of the ground state $|g\rangle$. Hence, after a finite time of propagation, the dot inevitably gets trapped in $|g\rangle$ and thereby forgets its initial state. So the stationary occupations are essentially determined by sequential tunneling alone. Even with all rates $W^{(1)}_{\phi' \leftarrow g}$ set to zero, the matrix $\boldsymbol{W}^{(1)}$ still has a

3. Influence of Spin Relaxation on Transport in Cotunneling Regime

rank of three and by Eq. (3.4) all $\mathbf{P}_\phi^{(0)}$ are fixed except for normalization.

In the core region sequential transitions out of both single-particle states are forbidden. Then classically the dot can get trapped either in the ground or in the excited state, so that the single-particle occupations depend on the initial dot state. Consequently, they are no longer determined by the lowest- but by the second-order rate equation, i.e., they are essentially given by cotunneling. This becomes manifest in the structure of $\boldsymbol{W}^{(1)}$, which has a rank of two, when all rates $W^{(1)}_{\phi' \leftarrow g,e}$ are set to zero. Between shell and core lies the intermediate region, where the system continuously changes between classical and cotunneling-dominated occupation, respectively. But as well as no matrix $\boldsymbol{W}^{(1)}$ can be constructed that continuously changes its rank, no single rate equation exists that both determines the systematic second-order expansion of $\mathbf{P}$ in terms of $\boldsymbol{W}^{(1)}$, $\boldsymbol{W}^{(2)}$ and is valid within all three regions simultaneously

Alternatively, we seek second-order approximations of $\mathbf{P}$ and I, which are valid in the cotunneling regime and perturbative in the sense that they deviate from the systematic expansions at most by terms quadratic and cubic in Γ, respectively. With the normalized solution $\mathbf{P}'$ of Eq. (3.4), in which $\boldsymbol{W}$ is replaced by the sum of the lowest two orders $\boldsymbol{W}^{(1+2)}$, Weymann et al. present an example for an approximation that is perturbative even for arbitrary values of eV_{bias} and $e\,\Phi_D^\epsilon$.[172] Unfortunately, for our system $\mathbf{P}'$ isn't well-defined within the entire core region, where the component P'_e becomes negative when $|eV_{\text{bias}}| < \Delta$ [green (dark gray), dotted area in Fig. 3.1(a)]. To resolve this problem we take into account that on the r.h.s. of the second-order equation[5]

$$\boldsymbol{W}^{(1)}\mathbf{P}^{(1)} = -\boldsymbol{W}^{(2)}\mathbf{P}^{(0)} \tag{3.7}$$

---

[5]Obtained by setting $n = 2$ in Eq. (3.6).

3. Influence of Spin Relaxation on Transport in Cotunneling Regime

the first-order probabilities $P^{(0)}_{0,d}$, which are exponentially small within the cotunneling regime, are multiplied with the rates $W^{(2)}_{\phi' \leftarrow 0,d}$. Hence, these rates drop out of Eq. (3.7) and its r.h.s. reduces to a vector V with components $V_{\phi'} = -\sum_\phi (\delta_{\phi,g} + \delta_{\phi,e}) W^{(2)}_{\phi' \leftarrow \phi}$. As a consequence the rates $W^{(2)}_{\phi' \leftarrow 0,d}$ don't contribute to systematic expansion orders given solely in terms of $W^{(1)}$ and $W^{(2)}$, and all contributions to P', they are contained in, are unsystematic and should be omitted. With regard to the approximation of the current the same is true for terms containing the rates $W^{L(2)}_{\phi' \leftarrow 0,d}$. By dropping the unsystematic terms we arrive at

$$0 = \left(W^{(1)} + \tilde{W}^{(2)}\right)\tilde{P} \tag{3.8}$$

and
$$\tilde{I} = -e \sum_\phi [(W^{L(1)} + \tilde{W}^{L(2)})\tilde{P}]_\phi, \tag{3.9}$$

where $\tilde{W}^{(2)}_{\phi' \leftarrow \phi} = (\delta_{\phi,g} + \delta_{\phi,e}) W^{(2)}_{\phi' \leftarrow \phi}$.

Without specifying a particular spin-flip mechanism, we include relaxation via an effective Hamiltonian

$$\hat{H}_{\rm rel} = \sum_q \left(\tau \hat{d}^\dagger_g \hat{b}^\dagger_q \hat{d}_e + h.c.\right), \tag{3.10}$$

which describes the coupling of the dot electrons to a bath of free particles with temperature T and $\hat{H}_{\rm bath} = \sum_q \epsilon_q \hat{b}^\dagger_q \hat{b}_q$ [6]. This coupling is characterized by the single complex parameter τ, giving the amplitude for a spin-flip process from $|e\rangle$ to $|g\rangle$. We assume, that the relaxation processes are completely incoherent to the electron tunneling and include only the first order of the perturbation expansion with respect to $\hat{H}_{\rm rel}$. Then, in the diagrammatic representation, the self-energy operator up to second order becomes the sum of all irreducible diagrams that have *either* up to two tunneling

---
[6] The letter q denotes a complete set of quantum numbers.

lines ($\hat{\Sigma}^{(1,2)}$) *or* exactly one relaxation line ($\hat{\Sigma}_{\text{rel}}^{(1)}$), which represents a Wick contraction of bath operators. The latter operator gives rise to an additional matrix term Θ in the master equation, whose matrix elements are defined in analogy to Eq. (3.3). Hence, we get the rate equation

$$0 = (\boldsymbol{W}^{(1)} + \tilde{\boldsymbol{W}}^{(2)} + \boldsymbol{\Theta})\tilde{\boldsymbol{P}} \qquad (3.11)$$

for a relaxation-dependent approximation $\tilde{\boldsymbol{P}}$. Evaluation of the relaxation diagrams then yields the rates

$$\begin{aligned}\Theta_{g\leftarrow e} = -\Theta_{e\leftarrow e} &= \frac{2\pi|\tau|^2}{\hbar}\int d\epsilon_q \langle \hat{b}_q \hat{b}_q^\dagger \rangle_{\text{b}} \rho_{\text{b}}(\epsilon_q)\delta(\epsilon_q - \Delta) \\ \Theta_{e\leftarrow g} = -\Theta_{g\leftarrow g} &= \frac{2\pi|\tau|^2}{\hbar}\int d\epsilon_q \langle \hat{b}_q^\dagger \hat{b}_q \rangle_{\text{b}} \rho_{\text{b}}(\epsilon_q)\delta(\epsilon_q - \Delta),\end{aligned} \qquad (3.12)$$

where $\rho_{\text{b}}(\epsilon_q)$ gives the density of states in the bath at energy ϵ_q and $\langle \cdot \rangle_{\text{b}}$ denotes the expectation value with respect to the bath degrees of freedom. Alternatively, these spin-flip rates can be calculated using standard time-dependent perturbation theory and Fermi's Golden rule (see, e.g., Ref. [184]). The first equalities in the Eqs. (3.12) express the conservation of the total probability, which in the diagrammatic approach is fulfilled by construction. Since we assume that the spin-splitting is large compared to the temperature ($\Delta\beta \gg 1$), the relaxation rates are approximately given by $\Theta_{\phi'\leftarrow\phi} = \delta_{\phi,e}(\delta_{\phi',g} - \delta_{\phi',e})\theta/\hbar$ with $\theta = 2\pi|\tau|^2 \rho_{\text{b}}(\Delta)$ both for a fermionic and a bosonic bath (as long as $|\mu_{\text{bath}}| \ll \Delta$).[7] $\tilde{\boldsymbol{P}}$ as well as $\tilde{I}$ are well-defined and perturbative within the cotunneling regime and seamlessly link in the intermediate region the systematic expansions that are only valid either in the core or shell. We also note that, because $\hat{H}_{\text{rel}}$ and $\hat{H}_r$ commute, the

---

[7]For a fermion bath with $\mu_{\text{bath}} = 0$ these rates lead to the same relaxation terms as in Eq. (2) of Ref. [55], if we identify $\theta = 2\hbar/\tau_{\text{sf}}$.

relaxation does not contribute (directly) to the current $\tilde{I}$, that is to say, Eq. (3.9) remains valid without modification.

3.3. Results

In this section we argue that the rich internal structure of the Coulomb diamond with its different overlapping regions and sub-regimes is responsible for the rather unexpected transport behavior the quantum dot shows in the presence of spin-relaxation. That is to say, the conductance peaks at the onset of sequential transport are, as stated above, maximally pronounced for a small finite relaxation rate. The peaks are situated close to the resonances with sequential transitions out of the excited state and therefore lie within the intermediate region. It turns out that the evolution of the peak height can be ascribed to the fact that in the core region of the Coulomb diamond the current $\tilde{I}(\theta)$ is much more sensitive to changes of the relaxation rate θ than it is in the shell. At small relaxation the current is diminished solely in the core region. Hence, the height of the current step is increased, while its width remains almost constant as compared to zero relaxation. It follows that the resulting conductance signatures in the intermediate region grow with the relaxation rate as long as the latter stays below a level at which the current in the shell region is affected.

At first we describe general features of electron transport through the SLQD before we explain in detail how it depends on the relaxation parameter θ. Fig. 3.1(b) shows a calculated charging diagram, i.e., the differential conductance $\tilde{G} = d\tilde{I}/dV_{\text{bias}}$ against eV_{bias} and $e\Phi_D^\epsilon$, of the cotunneling regime. The parameters are $\Delta = 45k_\text{B}T$, $U = 225k_\text{B}T$, $\Gamma = 4.5 \times 10^{-3} k_\text{B}T$ and $\theta = \Gamma/2$. The Coulomb diamond is defined by pronounced red (dark gray) lines of high conductance, where one of the reservoir's electro-chemical potentials is close to resonance with the energy of

3. Influence of Spin Relaxation on Transport in Cotunneling Regime

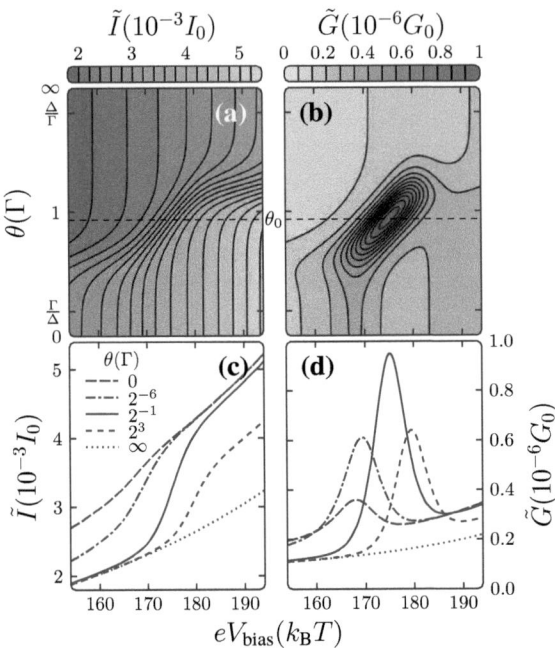

Figure 3.2.: Current $\tilde{I}$ (in units of $I_0 = e\Gamma/\hbar$) and differential conductance $\tilde{G}$ versus bias eV_{bias} and relaxation θ with $e\Phi_D^\epsilon = (U+\Delta)/2$ and parameters Δ, U, Γ as in Fig. 3.1. It can be seen how the height of the cotunneling-mediated current step (a) and of the corresponding conductance peak (b) as well as their positions depend on the relaxation rate. In the range of $0 \leq \theta \lesssim \Gamma^2/\Delta$ (linear scale) the system is hardly affected by the relaxation. Between Γ^2/Δ and Δ (logarithmic scale) the height of current step and conductance peak first grow to a maximum value at $\theta_0 \approx 0.58\Gamma$ (dashed lines in (a) and (b)) for increasing θ, then decrease again and vanish before $\theta = \Delta$. Even faster relaxation (reciprocal scale) has no further effect. Figures (c) and (d) show cuts through (a) and (b), respectively, for five different values of θ. Both the step in (a,c) and peak in (b,d) slightly shift towards higher absolute values of eV_{bias} for increasing rates between Γ^2/Δ and Δ.

3. Influence of Spin Relaxation on Transport in Cotunneling Regime

a single-charge transition involving the ground state. Due to the Coulomb blockade the slightly weaker resonance lines of excited state transitions are not extended into the diamond. Instead, thin red (dark gray) lines appear in that part of the intermediate region, in which $|eV_{\text{bias}}| > \Delta$, corresponding to the yellow (light gray), not dotted area in Fig. 3.1(a). [53, 84] These are the signatures that mark the onset of cotunneling-mediated sequential transport out of the excited state. The light vertical shades along the lines $eV_{\text{bias}} = \pm\Delta$ arise from the above-mentioned opening of inelastic transport channels. Though being actually invalid near the border of the Coulomb diamond and in its exterior, the approximate solution $\tilde{I}$ was used for Fig. 3.1(b) even in these regions, because its deviation from the systematic solution $I^{(1+2)}$ turned out—for the chosen set of parameters—to be too small to be visible. In general it may be necessary to use the systematic expansion for the border and outer region, which can be seamlessly connected to the approximate solution in the shell region given that $\Gamma\beta \ll 1$[8].

The dependence of current and conductance on the relaxation rate θ in the vicinity of the excited state resonance is shown in Fig. 3.2 part (a) and (b), respectively, [the range of bias values eV_{bias} is marked by the short horizontal line in Fig. 3.1(b)]. The parameters Δ, U, and Γ are the same as in Fig. 3.1 and $e\,\Phi_D^\epsilon = (U+\Delta)/2$. If the rate θ is smaller than Γ^2/Δ, thus well below the lowest second-order tunneling rate, the relaxation hardly affects $\tilde{I}$ and $\tilde{G}$, so that the system behaves as for $\theta = 0$. As the rates grow between Γ^2/Δ and $\theta_0 \approx 0.58\Gamma$ they eventually become much larger than every cotunneling rate, while still being smaller than sequential rates of energetically allowed processes. Height and slope of the current step as well as the height of the conductance peak increase in this range to a maximum value at $\theta = \theta_0$, since the relaxation diminishes $\tilde{I}$ in the inelastic

---

[8] Of course, if Δ is too small to lead to a sufficiently broad shell region, no (smooth) crossover can be found.

3. Influence of Spin Relaxation on Transport in Cotunneling Regime

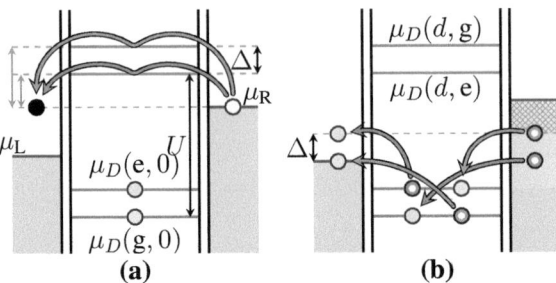

Figure 3.3.: Position of the Fermi levels μ_L and μ_R of left and right reservoir (blue (very dark gray)) relative to the four energy differences $\mu_D(\phi', \phi) = \epsilon_{\phi'} - \epsilon_\phi - e\Phi_D$ (light red (light gray) horizontal lines) between initial and finial state energy of the single-particle transitions (parameters $\Delta, U, \Phi_D^\epsilon$ as in Fig. 3.1). In (a) bold green (light gray) and red (dark grey) arrows represent elastic tunneling via virtual state $|d\rangle$ through the dot in ground and excited state, respectively, indicated by filled green (light gray) and red (dark gray) points ($eV_{\text{bias}} = 1.5\Delta$). Double pointed arrows on the left show for both cases how much energy the tunneling electron must gain to get into the virtual state. (b) Coherent inelastic processes causing transitions $|g\rangle \to |e\rangle$ (green (light gray) arrows) and $|e\rangle \to |g\rangle$ (red (dark gray) arrows) are illustrated ($eV_{\text{bias}} = 2\Delta$). Electrons in the red (dark gray) colored (green (light gray) hatched) part of reservoir R can tunnel into the dot, if it is in initial state $|e\rangle$ ($|g\rangle$).

part of the core and low bias part of the intermediate region but leaves it almost unaltered in the sequential part of the shell region. A relaxation with rate $\theta_0 \leq \theta \leq \Delta$ is (roughly) as fast as or faster than sequential tunneling and while $\tilde{I}$ in this parameter range shows no further relaxation dependence in the core, it decreases in the shell region for growing θ. As a result the current step and conductance peak decrease as well and vanish before $\theta = \Delta \gg \Gamma$. For rates $\Delta \leq \theta \leq \infty$ transport properties do not depend on θ. Obviously there exists an optimal relaxation rate θ_0 for which the signatures of cotunneling-mediated sequential transport have maximal height and are

3. Influence of Spin Relaxation on Transport in Cotunneling Regime

considerably more pronounced than for $\theta = 0$. Also do the positions of the resonance signatures move to higher absolute values of eV_{bias}, when θ is increased between Γ^2/Δ and Δ. As θ_0 depends on the ratio Δ/U, the coupling Γ, and temperature β in a complicated way, we can give no simple estimation of its value in terms of the system parameters other than: $\theta_0 = c\,\Gamma$ with $0.3 \lesssim c \lesssim 1.2$ for all parameter sets yielding reliable results. The height of the conductance peaks depends most strongly on θ throughout the whole range $(\Gamma/\Delta)^{1/2} \lesssim \theta/\Gamma \lesssim (\Delta/\Gamma)^{1/2}$ with $\Gamma \ll \Delta$ [see Fig. 3.2(b)], while its relative variation for all values of c amounts to only a few percent. Therefore, we content ourselves with the statement, that θ_0 corresponds to a rate that is roughly as large as Γ and very much larger than any cotunneling rate. In order to explain these observations, we examine how the tunneling processes that dominate the current in the relevant parts of the cotunneling regime are influenced by the relaxation.

In the inelastic part of the core region the current is caused solely by elastic and inelastic cotunneling, which also dominates the occupation of the single-particle states. In particular, inelastic tunneling provides an occupation of the excited state of order 1 that doesn't depend on Γ and is reduced by the relaxation, as soon as the magnitude of θ becomes at least comparable to $W_{e \leftarrow g}^{(2)}$. Since the $\tilde{P}_{0,d}$ are much smaller than the single-particle occupations in the cotunneling regime, the current consists mainly of two contributions associated with cotunneling out of state $|g\rangle$ and $|e\rangle$, which are proportional to $\tilde{P}_g$ and $\tilde{P}_e$, respectively. Hence, the cotunneling based on processes with initial state $|g\rangle$ benefits from a change $d\tilde{P}_g(\theta) \approx -d\tilde{P}_e(\theta) > 0$ caused by relaxation, whereas the current with the dot being initially in state $|e\rangle$ is decreased by it. So in the core region the dependence of the current change $d\tilde{I}$ on $d\tilde{P}_g(\theta)$ is given by $d\tilde{I}/d\tilde{P}_g(\theta) \approx -e/\hbar \sum_{\phi'} \left(W_{\phi' \leftarrow g}^{L,(2)} - W_{\phi' \leftarrow e}^{L,(2)} \right)$. The sums on the r.h.s. have the same sign, which is for both contributions specified by the direction of current flow and thus by the sign of V_{bias}. For the

3. Influence of Spin Relaxation on Transport in Cotunneling Regime

inelastic part of the core one can establish the relation

$$\left|\sum_{\phi'} W^{L,(2)}_{\phi'\leftarrow e}\right| > \left|\sum_{\phi'} W^{L,(2)}_{\phi'\leftarrow g}\right| \qquad (3.13)$$

by looking at the energy dependence of elastic and inelastic processes. In the core, an electron that elastically tunnels through the dot with initial (and final) state $|\chi = g, e\rangle$ via virtual intermediate state $|d\rangle$ has to overcome at least the energy difference $U + \delta_{\chi,g}\Delta - e(\Phi^\epsilon_D + |V_{\text{bias}}|/2) > 0$, which is by Δ smaller for an initially excited dot than for one in the ground state (s. Fig. 3.3 (a)). The latter is also true for tunneling via virtual state $|0\rangle$, which can be seen analogously. As a consequence the rate for elastic cotunneling is smaller for $\chi = g$ than for $\chi = e$.

For the inelastic processes a similar energy argument can be applied. Inelastic tunneling out of the ground into the excited state cannot set in before $e|V_{\text{bias}}| = \Delta$, because the energy Δ, needed for the transition to take place, has to be provided by the reservoirs. In contrast, inelastic tunneling, causing the opposite transition, is always possible, because in this case the transition energy is provided by the dot. Hence, if V_{bias} and Φ^ϵ_D specify a point in the core region, for $\chi = e$ there are always more electrons available for inelastic processes compared to the case $\chi = g$ (s. Fig. 3.3 (b)). This results in a higher rate for inelastic tunneling through a dot in initial state $|e\rangle$ and immediately leads to Eqn. (3.13). Using this equation and the fact that $d\tilde{P}_g(\theta)$ is a positive, monotonic function in θ, for the core we can derive

$$\frac{d|\tilde{I}|}{d\theta} = \frac{e}{\hbar}\frac{d\tilde{P}_g}{d\theta}\left(\left|\sum_{\phi'} W^{L,(2)}_{\phi'\leftarrow g}\right| - \left|\sum_{\phi'} W^{L,(2)}_{\phi'\leftarrow e}\right|\right) \leq 0. \qquad (3.14)$$

Since in this region the single-particle occupations are determined mainly by cotunneling, they only depend on the relaxation, if θ is comparable to the second-order rates. Due to the factor $d\tilde{P}_g/d\theta$ on the r.h.s. of Eqn. (3.14), this dependence also holds for $\tilde{I}$, whose absolute value decreases for θ growing between Γ^2/Δ and Γ and is constant for slower or faster relaxation, respectively.

In the inelastic part of the shell region the maximal cotunneling-provided occupation of the excited state is by a factor Γ smaller than in the inelastic core, since it is reduced by sequential transport out of state $|e\rangle$, as long as $\theta \lesssim W_{0,d\leftarrow e}^{L,(1)} \approx \Gamma$. When the relaxation becomes faster than sequential tunneling, $\tilde{P}_e$ decreases and eventually goes to 0 for $\theta \gg \Gamma$. The difference in size of sequential and cotunneling rates compensates for the reduction of the excited state occupation, so that the total current is higher in the shell compared to the core region. It consists of contributions associated with sequential tunneling out of states $|\chi = 0, e, d\rangle$ and cotunneling out of the ground state.

When θ is not much larger than Γ these contributions are all of the same order of magnitude, which is, however, not the case for their response to increasing relaxation. Obviously, relaxation rates much higher than Γ completely depopulate the excited state and the current caused by tunneling out of $|e\rangle$ vanishes. Its relative change in magnitude compared to the case of low relaxation is therefore of order 1. For cotunneling out of state $|g\rangle$, on the other hand, the maximum relative change is

$$\frac{\tilde{P}_g(\theta \gg \Gamma) - \tilde{P}_g(\theta \ll \Gamma)}{\tilde{P}_g(\theta \ll \Gamma)} = \frac{\mathcal{O}(\Gamma)}{(1 - \mathcal{O}(\Gamma))} \approx \mathcal{O}(\Gamma). \tag{3.15}$$

Hence, if the relaxation increases, the gain in the cotunneling current associated with $\tilde{P}_g$ cannot compensate for the simultaneous suppression of the sequential current proportional to $\tilde{P}_e$, which results in a reduced total current. Similarly to the discussion of the core, however, it can be argued that

3. Influence of Spin Relaxation on Transport in Cotunneling Regime

in the shell, where the single-particle occupations are mainly determined by sequential processes, the total current can only show a considerable relaxation dependence, if θ is neither much smaller nor much larger than Γ or, equivalently, than the rates for sequential tunneling. As we stated above, the current dependence on θ smoothly crosses over between the core- and shell-like behavior in the intermediate region, so that both the current step and the conductance peak grow with θ between Γ^2/Δ and $\theta_0 \approx c\,\Gamma$, while they decrease for $\theta > \theta_0$ and vanish before $\theta = \Delta$. The fact that the current becomes less sensitive to relaxation for higher values of $e\,|V_{\text{bias}}|$, showing a sharp step-like dependence in the intermediate region, also manifests in the slight shift of the position of the excited state resonances to higher absolute bias values.

3.4. Summary

In this chapter we discussed Coulomb-blocked electron transport through a SLQD with spin-split level that is coupled to two non-magnetic, metallic leads. We used the real-time diagrammatic technique to systematically expand occupation probabilities and tunneling current up to the second-order in the strength Γ of the tunnel coupling, thereby including sequential and cotunneling into the transport calculations. Two properties were considered with respect to which the Coulomb blockade regime can be subdivided into parts that differ at least in one of them: the kind of tunneling processes contributing to the current (elastic, inelastic, sequential transport) and those determining the single-particle occupations (cotunneling in the core, sequential tunneling in the shell region). At or close to the borders between these sub-regimes, signatures of the dot's excitation spectrum appear in the current and differential conductance. With the focus on excited state signatures marking the onset of cotunneling-mediated sequential transport, we studied

3. Influence of Spin Relaxation on Transport in Cotunneling Regime

how the current is influenced by a phenomenologically introduced spin relaxation with rate θ. It turned out that for a relaxation rate of about half the tunnel coupling the excited state resonances are maximally pronounced, being considerably larger than without relaxation, while in the limit of infinite θ the resonances completely vanish. We explained this behavior by a combination of two effects: (i) the current decreases monotonically with growing relaxation rates and (ii) the excited state occupation is in the cotunneling-dominated core and in the shell region only affected by a relaxation with rates in the range $\Gamma^2/\Delta < \theta \lesssim c\Gamma$ and $c\Gamma \lesssim \theta < \Delta$ with $0.3 \lesssim c \lesssim 1.2$, respectively.

This relaxation dependence of the current may illuminate why the resonance signatures measured in Ref. [84] are relatively sharp compared to the ones that were calculated for $\theta = 0$. Furthermore it could provide means to directly influence the single-particle occupations in experiments and allows to facilitate measurements of excited state resonances by adjusting either the coupling Γ or the rate θ.

4

Iterative Summation of Path Integrals

IT WAS SHOWN in the previous chapters, that many fundamental physics and principal features of transport through a SLQD can be explained both efficiently and intuitively within well-established, perturbative frameworks. At least this is true for cases, in which an energy scale in the system is small enough to be considered as weak perturbation (of an otherwise solved problem) and, yet, related to a source of qualitatively new behaviour. As an important example of this case, we discussed the limit of weak tunnel coupling so far, a regime, which is also accessible by state-of-the-art experiments. Without the small coupling, no tunnelling current would flow. On the other hand, even the smallest contact will result in a stochastic dynamics caused by a multitude of fluctuating tunnelling events and, depending on the parameters, a drastic change of the dot's (stationary) state. Not only are perturbative approaches often in good agreement with experimental data. Their considerable explanatory power is to no small part due to the intuitive picture of physical processes they provide.

Despite being a valuable and versatile tool (not only) in the theory of mesoscopic transport, however, the applicability of perturbation theory is

4. Iterative Summation of Path Integrals

limited by the need for a small parameter and there is physics that lie beyond its means. Obvious counter-examples involve systems, in which the aforementioned condition does not hold, i.e., all appearing energy scales are of the same order of magnitude. This is, in fact, the kind of situation, we want to study in this and the next chapter. But besides systems that are found by this rather quantitative criterion, others may withstand a perturbative analysis, even when a small parameter is present. Prominent examples include superconductivity [185] and the Kondo effect [135, 136], both of which appear at low temperatures. What all these cases have in common is a strong significance of coherent (higher-order) processes, which determine the system's observed properties, i.e., no simple relation between a single (small) energy parameter and the altering effect on the system can be found, which quantifies the contribution of expansion terms in orders of this parameter. From a fundamental point of view, this is interesting, since coherence effects are a characteristic feature of quantum mechanics. Systems with these as an essential constituent thus grant a direct glance into the quantum world. Yet, quantum effects and coherence are also promising ingredients for novel applications and devices that show properties unknown in conventional electronics. Specifically, for our model of a SLQD with magnetic impurity, it is the interaction with and manipulation of a quantum spin by a conventional (unpolarised) charge current, that is of interest. Specifically, we want to explore the transport regime, where the tunnel coupling between dot and leads is comparable (in size) to the dot energies and where lead induced correlations are expected to be as important as the coherent on-dot dynamics.

Based on the principles and concepts we laid down in chapter 2, a fermionic path integral representation of the Keldysh generating function $\mathcal{Z}[\eta]$ is derived in the following section. Subsequently, the approach of the *iterative summation of path integrals* (ISPI) is introduced in section 4.2. It

4. Iterative Summation of Path Integrals

was developed by Weiss et al. [120] and as we follow, in large parts, the derivation shown in this paper, we also show how to adopt the method to model (2.1) with the fixed spin-1/2 magnetic impurity. The chapter closes with a discussion of the numerical complexity of ISPI and how it can, in certain circumstances, be significantly reduced by restricting the path sum to certain classes of impurity spin paths.

4.1. Generating Function and Fermionic Path Integral

Conceptional foundation of the derivations in this chapter is the Keldysh partition function $\mathcal{Z}$ (2.11). We showed at the end of section 2.2.2 how a generating function $\mathcal{Z}[\eta]$, which is derived from the partition function, can be used to calculate expectation values of arbitrary observables [see equation (2.13)]. The ISPI method builds on this relation, for it employs a path integral representation of $\mathcal{Z}[\eta]$ as starting point of all numerical calculations. The *electronic* part of the model that is studied in this thesis requires a particular representation based upon the *fermionic path integral* [86, 186, 187] and the *Grassmann algebra*.

All matters concerning the time evolution of a quantum system can in principle be broken down to one kind of elementary question. What is the probability amplitude to find a quantum system in Hilbert state $|\psi_f\rangle$ at $t_f > t_i$, when it was initially in state $|\psi_i\rangle$ at time t_i? In this context, the evolution operator $\hat{U}(t, t')$ [see equation (2.5)], which is described in section 2.2.2 and appendix A, plays the central role. It contains in its action the complete dynamics as given by the system Hamiltonian. Hence, with the help of $\hat{U}$,

4. Iterative Summation of Path Integrals

the sought-after amplitude can be identified with matrix element

$$U(f|i) := \langle \psi_f | \hat{U}(t_f, t_i) | \psi_i \rangle. \tag{4.1}$$

Therefore, as a first step towards a path integral representation of $\mathcal{Z}[\eta]$, we construct a corresponding expression for the evolution operator. This is done by splitting the propagation time $t_f - t_i$ into a large number $N-1$ of short time spans $\delta_t := (t_f - t_i)/(N-1)$, where the short time propagation is described by a comparably simple approximate propagator, whose proper construction depends on the particular model system.

4.1.1. The Short Time Propagator

To this end, we have to consider the limit of "very short" propagation times δ_t in detail. Our aim is to find a suitable *short time propagator* $\hat{U}_{\delta_t}(t)$, defined by the relation

$$\hat{U}(t + \delta_t, t) = \hat{U}_{\delta_t}(t) + \mathcal{O}(\delta_t^2). \tag{4.2}$$

In other words, a short time evolution operator equals the full propagator up to terms scaling quadratically in the duration δ_t of the short time span between t and $t + \delta_t$. This definition does not uniquely determine a particular $\hat{U}_{\delta_t}(t)$, but rather a whole class of operators. The generality of this relation grants a certain amount of flexibility, as to the choice of the particular form of short time evolution operator. In the following this flexibility is used to construct a variant most suitable to model (2.1). A time span shall be regarded as "very short", if

(4.S1) δ_t it is much smaller than all dynamical time scales in the system, as given by the energy scales ϵ in the Hamiltonian and a possible time dependence of the Hamiltonian itself: $\delta_t \ll \hbar/\epsilon$.

4. Iterative Summation of Path Integrals

(4.S2) The system can undergo, during δ_t, *at most one* transition between states that are connected via an *elementary process*.

The possible elementary state transitions (processes) are given by the interaction or hybridisation terms in the Hamiltonian. For model (2.1), examples are a single-electron tunnelling event due to $\hat{H}_T$ and a flip-flop process of electron- and impurity spin due to the second term on the r.h.s. of equation (2.1c). The time scale of these transitions themselves, however, is infinitely smaller than δ_t. This is suggested by the qualifier "elementary", as referring to *instantaneous* transitions with no time structure, and corresponds to the limit of infinitesimal (infinitely small) duration dt. These abstract elementary transitions, however, do not (directly) describe physically observable processes like, e.g., flip-flops. Rather they are a conceptual tool for the construction and physical interpretation of quantum dynamics within the path integral formulation. The mathematical relationship between elementary state transitions and the corresponding physical processes will be shown in the following.

For an intuitive illustration of this relationship, we consider the free precessing motion of a quantum spin-1/2 in the perpendicular magnetic field $\boldsymbol{B} = B\boldsymbol{e}_x$. In this system, elementary state transitions are given by terms proportional to the Pauli matrices $\hat{\tau}_\pm$, which describe spin flips between states $|\tau =\uparrow,\downarrow\rangle$. If the physical propagation of the spin was given, as suggested above, by a sequence of periods of free propagation interrupted by instantaneous, elementary transitions, the time-dependent orientation of a spin that was initially in state $|\uparrow\rangle$, for example, would be a piecewise continuous, step-like function. In reality, however, the spin orientation is a smooth function that oscillates with a frequency proportional to the strength B of the magnetic field [see Fig. 4.1(a)]. Though individual step-like paths in state space (solid lines on the l.h.s.) describe an unphysical time evolution,

4. Iterative Summation of Path Integrals

the *path integral*, which takes into account all possible paths of this kind, yields the correct physics. As we show below, each path contributes a phase factor determined by its classical action. Hence, the path integral formulation can be interpreted in the following way: a quantum system/particle travels *simultaneously* along *all possible paths* (in state space) that involve free propagation and elementary transitions only [see Fig. 4.1(a)].

With the assumptions (4.S1) and (4.S2), the structure of a short time evolution operator can be deduced from the full propagator (2.5). For a Hamiltonian $\hat{H}(t)$ and short times δ_t, we can write

$$\hat{U}(t + \delta_t, t) = e^{-i/\hbar \hat{H}(t)\delta_t} = \sum_{n=0}^{\infty} \frac{(-i\delta_t/\hbar)^n}{n!} \hat{H}^n(t)$$

$$= \hat{\mathbb{1}} - \frac{i}{\hbar}\hat{H}(t)\delta_t + \mathcal{O}(\delta_t^2).$$
(4.3)

Hence, any operator, whose two lowest-order terms equal those of (4.3), qualifies as short time propagator. In fact, we could just use $\hat{U}_{\delta_t}(t) := \hat{\mathbb{1}} - i/\hbar \hat{H}(t)\delta_t$ to construct the path integral. Another valid choice would be the normal-ordered evolution operator

$$:\exp\left\{-\frac{i}{\hbar}\hat{H}(t)\delta_t\right\}: = \sum_{n=0}^{\infty} \frac{(-i\delta_t/\hbar)^n}{n!} :\hat{H}^n(t):,$$
(4.4)

where in the normal-ordered form (or normal form) $:\hat{O}:$ of an operator $\hat{O}$, creation operators appear left of annihilation operators. The main relevance of version (4.4) lies in the boundedness for bosonic Hamiltonians in the whole complex time plane [186], but it is also convenient to use, when—as in the present work—the formulation of the path integral is based on coherent states (see below).

4. Iterative Summation of Path Integrals

Figure 4.1.: The relationship between the abstract concept of elementary state transitions and the real-time dynamics of a physical system, illustrated with the help a spin-1/2 in a magnetic field perpendicular to its quantisation axis ($B = Be_x$). **(a)** If the spin was initially in the state $|\uparrow\rangle$, the physical evolution of the spin polarisation $\langle\tau_z\rangle$ (dashed line) shows a smooth, oscillating behaviour (with a frequency that is proportional to B). A time evolution described by a sequence of free propagation periods and spin flips (solid line) would give an unphysical, step-like functional dependence of the polarisation. **(b)** Though individual step-like paths in state space (solid lines on the l.h.s.) do not correctly describe the spin's physical evolution, the *path integral*, which takes into account all possible paths of this kind, yields the correct physics. Each path contributes a phase factor determined by its classical action (see below).

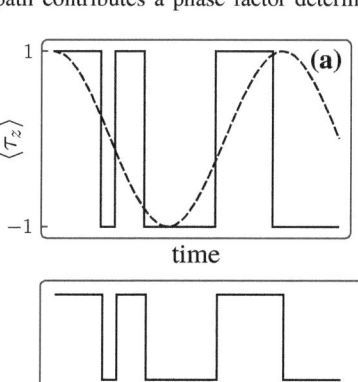

According the path integral interpretation, the spin travels along *all those paths simultaneously* and elementary state transitions cannot be interpreted as real physical processes. Rather, the are a conceptual tool used to construct and interpret the path integral.

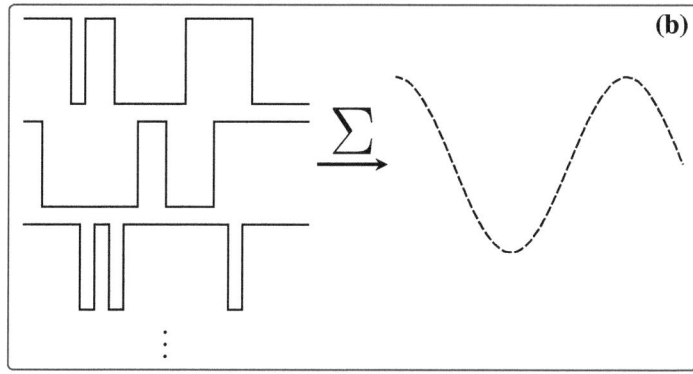

4. Iterative Summation of Path Integrals

For the model with impurity interaction and Coulomb repulsion, both of these two examples for short time propagators, though valid, are not the optimal choice with regard to clearness and simplicity of the following derivations. The reason is, that the numerical ISPI approach is built upon a path integral expression, in which the two-particle interactions (Coulomb and electron-impurity) manifest themselves in sums over tuples (paths) of discrete state variables[1] rather than integrals over continuous paths. For the Coulomb interaction, this is achieved by a so-called Hubbard-Stratonovich transformation. The same procedure may very well be applicable to the impurity interaction as well, if the impurity spin is described by spin coherent states [188–190] or in a semi-fermionic framework [191, 192]. Yet, using the complete basis of states $|\tau = \pm 1\rangle$ to represent the impurity will turn out to be the easier way. On the downside, it will require to deal with a "mixed" basis, namely the tensor product of fermionic coherent states for electrons and the discrete spin states for the impurity. Besides the Coulomb interaction, it is the use of this basis, which leads to a slightly more complicated form of $\hat{U}_{\delta_t}(t)$. But once the appropriate short time propagator is found, the derivation of the path integral goes the usual way for fermionic systems. In the remainder of this section, we assume stationarity of $\hat{H}$ for convenience. The general, time-dependent results are easily obtained from the final expressions by re-introducing the time index to the respective Hamiltonian terms.

To construct a proper $\hat{U}_{\delta_t}$, it is helpful to separate in $\hat{H} = \hat{H}_0 + \hat{H}_{\text{TR}}$ terms $\hat{H}_0$ that are diagonal in the non-interacting many-particle basis $\{|\psi\rangle\}$ from those terms $\hat{H}_{\text{TR}}$ that cause elementary transitions. For the model

---

[1] The spin orientation $\tau = \pm 1$, e.g., in case of the impurity.

4. Iterative Summation of Path Integrals

Hamiltonian (2.1), we have

$$\hat{H}_0 = \overbrace{\sum_\sigma \epsilon_\sigma \hat{n}_\sigma + \hat{H}_{\text{imp}}}^{\hat{H}^0_{\text{dot}}} + \overbrace{J \sum_\sigma \sigma \hat{\tau}_z \hat{n}_\sigma}^{\hat{H}^\parallel_{\text{int}}} + \hat{H}_{\text{leads}} \qquad (4.5a)$$

$$\hat{H}_{\text{TR}} = \underbrace{U \hat{n}_\uparrow \hat{n}_\downarrow}_{\hat{H}^U_{\text{dot}}} + \underbrace{J \sum_\sigma \hat{\tau}_\sigma \hat{d}^\dagger_{-\sigma} \hat{d}_\sigma / 2}_{\hat{H}^\perp_{\text{int}}} + \hat{H}_T, \qquad (4.5b)$$

with $\hat{H}^0_{\text{dot}} + \hat{H}^U_{\text{dot}} = \hat{H}^{\text{el}}_{\text{dot}}$ and $\hat{H}^\parallel_{\text{int}} + \hat{H}^\perp_{\text{int}} = \hat{H}_{\text{int}}$. If the Hamiltonian consisted only of diagonal parts, the effect of acting with the corresponding evolution operator $\hat{U}_0(\Delta_t) := \exp\{-i/\hbar \hat{H}_0 \Delta_t\}$ for time interval $\Delta_t := t_f - t_i$ on a non-interacting state $|\psi\rangle$ would amount to a multiplication with phase factor $\exp\{-iE_{0,\psi}\Delta_t/\hbar\}$, where $E_{0,\psi} = \langle \psi | \hat{H}_0 | \psi \rangle$ is the (free) energy of state $|\psi\rangle$. In other words, a state $|\psi\rangle$ would propagate freely. Transitions and interactions in the system enter via $\hat{H}_{\text{TR}}$. For simplicity, we call all events caused by any of its parts "transitions" in the following paragraphs.

In this context, a representation in the interaction picture proves to be helpful establishing a suitable version of the short time propagator, as it can be applied to any decomposition of the Hamiltonian into two terms $\hat{H} = \hat{H}_0 + \hat{H}_1$. We start with the full time propagator in the interaction picture (see appendix C)

$$\hat{U}(t_f, t_i) = \sum_{N=2}^{\infty} \int_{t_i}^{t_f} \cdots \int_{t_{N-2}}^{t_f} \hat{U}_0(\Delta_t^{(N-1)}) \left(-\frac{i}{\hbar} \hat{H}_1 dt_{N-1}\right) \cdots \hat{U}_0(\Delta_t^{(2)}) \times$$

$$\cdots \times \left(-\frac{i}{\hbar} \hat{H}_1 dt_2\right) \hat{U}_0(\Delta_t^{(1)}) \qquad (4.6)$$

$$= \sum_{N=2}^{\infty} \int_{t_i}^{t_f} \cdots \int_{t_{N-2}}^{t_f} \underset{t_i}{\bullet}\!-\!\boxed{\hat{H}_1}\!-\!\boxed{\hat{H}_1}\!-\!\boxed{\hat{H}_1}\!-\cdots-\!\boxed{\hat{H}_1}\!-\!\underset{t_f}{\bullet}.$$
$$ t_2 \quad\;\; t_3 \quad\;\; t_4 \qquad\;\; t_{N-1}$$

4. Iterative Summation of Path Integrals

Here, we defined time differences $\Delta_t^{(k)} := t_{k+1} - t_k$ with $t_1 := t_i$ and $t_N := t_f$ as well as the graphical notations

$$\hat{U}_0(\Delta_t) =: \underset{t_i \quad t_f}{\bullet\!\!-\!\!\!-\!\!\!\rightarrow\!\!\bullet} \quad \text{and} \quad -\frac{i}{\hbar}\hat{H}_1 dt =: \boxed{\hat{H}_1}. \qquad (4.7)$$

This definition is chosen in such a way that for each N the contour is broken down into $N - 1$ pieces of free propagation that are connected by $N - 2$ interaction vertices.

With $\hat{H}_0$ defined in equation (4.5), we could identify $\hat{H}_1 \equiv \hat{H}_{TR}$. For reasons that will become clear below, we do not fix a particular $\hat{H}_1$ at this point; it turns out, that it is convenient to derive the final expression for $\hat{U}_{\delta_t}$ by adding the three interaction terms in $\hat{H}_{TR}$ one by one to $\hat{H}_0$. Rather, we just assume, that $\hat{H}_1$ describes elementary transitions between the Hilbert states $|\psi\rangle$. The free propagator is represented by a line segment, while $\hat{H}_1$ is depicted as transition vertex (box). Each vertex comes with an additional factor $-i/\hbar$ and dt, the latter of which can be interpreted as the time duration of the elementary transition. Thus, the full evolution operator is the sum of all diagrams with $N - 2 = 0, 1, \ldots$ transition vertices at times $t_2, \ldots, t_{N-1}$. Over all these vertex times between t_i and t_f is integrated, while preserving the order $t_1 < t_2 < \ldots < t_{N-1} < t_N$. To increase the ease of reading, we will skip the time integrals in front of the diagrammatic expressions in the following.

According to (4.S1) and (4.S2), the system can either propagate freely during small time δ_t or undergo exactly one elementary transition at some

4. Iterative Summation of Path Integrals

time $0 < t < \delta_t$. Hence the short time propagator can be written as

$$\hat{U}_{\delta_t} = \underset{0\quad\quad\delta_t}{\bullet\!\!\longrightarrow\!\!\bullet} + \underset{0\quad\quad\quad\delta_t}{\bullet\!\!\longrightarrow\!\!\boxed{\hat{H}_1}\!\!\longrightarrow\!\!\bullet}$$

$$= \hat{U}_0(\delta_t) - \frac{i}{\hbar}\int_0^{\delta_t} dt\, \hat{U}_0(\delta_t - t)\hat{H}_1\hat{U}_0(t) \quad (4.8)$$

$$= :\hat{U}_0(\delta_t)\left(\hat{\mathbb{1}} - \frac{i}{\hbar}\hat{H}_1\delta_t\right): + \mathcal{O}(\delta_t^2),$$

where the commutator relation $[\hat{U}_0(t), \hat{H}_1] = \mathcal{O}(t)$ was used to pull $\hat{H}_1$ through to the right of propagator $\hat{U}_0(t)$. The remaining time integration can then be carried out to yield an expression that is exact up to terms $\mathcal{O}(\delta_t^2)$. An error of the same size is obtained by the normal ordering. The definition of normal order of the free evolution part only refers to the electronic operators; since the free impurity Hamiltonian $\hat{H}_{\text{imp}}$ commutes with the remaining terms in $\hat{H}_0$, its position is arbitrary. Of course, due to the presence of $\hat{H}_{\text{int}}$, this is not true for the full Hamiltonian and the corresponding evolution operator. Hence, in later steps of our derivation, the relative order of non-commuting impurity operators has to be taken into account.

Once acquired, the short time propagator is used to break up a time evolution that extends over an arbitrary time span from t_i to t_f into $N-1$ repeated propagations of duration δ_t, supposed that $(N-1)\delta_t = t_f - t_i$. This so-called *Trotter break-up* of the time contour [193–195] yields, when plugged into equation (4.1), an approximation for the matrix element of the full propagator

$$U(f|i) = \langle\psi_f| \prod_{k=1}^{N-1} \hat{U}_{\delta_t}(t_i + [k-1]\delta_t)|\psi_i\rangle + \mathcal{O}(\delta_t^2), \quad (4.9)$$

which is again exact up to second-order terms in δ_t. For a stationary Hamiltonian, the operator product becomes the power $\hat{U}_{\delta_t}^{N-1}$. Finally, by inserting

4. Iterative Summation of Path Integrals

a partition of unity—here $\hat{\mathbb{1}} = \sum_\psi |\psi\rangle\langle\psi|$, for example—between every two short time propagators $\hat{U}_{\delta_t}$ in (4.9) we arrive at the path integral (or rather path sum)

$$U(f|i) \approx \sum_{\psi_2\ldots\psi_{N-1}} \langle\psi_f|\hat{U}_{\delta_t}|\psi_{N-1}\rangle \cdots \langle\psi_3|\hat{U}_{\delta_t}|\psi_2\rangle\langle\psi_2|\hat{U}_{\delta_t}|\psi_i\rangle. \quad (4.10)$$

The matrix element is approximated as a sum over all tuples or paths $(\psi_{N-1}, \ldots, \psi_2)$ of state indices, where each tuple contributes a factor $\mathcal{U} := \prod_{i=1}^{N-1} \langle\psi_{i+1}|\hat{U}_{\delta_t}|\psi_i\rangle$ with fixing $\psi_1 := \psi_i$ and $\psi_N := \psi_f$. A path contribution is non-zero, if none of the matrix elements $\langle\psi_{i+1}|\hat{U}_{\delta_t}|\psi_i\rangle$ vanishes. With the structure of $\hat{U}_{\delta_t}$ in mind, it follows, that this is the case if either $|\psi_i\rangle$ and $|\psi_{i+1}\rangle$ are equal (free evolution) or they are connected by an elementary transition. A tuple $\Psi := (\psi_N, \ldots, \psi_1)$ that contributes to (4.10) can thus be interpreted as a possible (time discrete) path the system can follow in the Hilbert space to get from $|\psi_i\rangle$ to $|\psi_f\rangle$ by a sequence of free evolutions and elementary transitions. The path integral is the sum over all such paths Ψ weighted with the product $\mathcal{U}(\Psi)$ of short time evolution matrix elements for every step δ_t. The limit $\delta_t \to 0$ of this path sum converges to the exact value

$$U(f|i) = \lim_{\delta_t \to 0} \sum_\Psi \Phi(\Psi) =: \int \mathcal{D}[\psi(t)]\,\mathcal{U}[\psi(t)], \quad (4.11)$$

which is often written in a continuous form as defined on the r.h.s.. This *notation* suggests, that the exact matrix element can be calculated as the integral over all continuous paths $\psi(t)$ with $t_i \leq t \leq t_f$ and $\psi(t_i) = \psi_i$, $\psi(t_f) = \psi_f$. Yet, a corresponding measure $\mathcal{D}[\psi]$ for such an integration does not exist, which is why the last step must not be understood as a mathematical identity relation but only as definition of a symbolic notation.

Nevertheless, though it has to be handled with care, this notation is widely used for reasons of clearness and simplicity; it complies to physical in-

4. Iterative Summation of Path Integrals

tuition to describe observables, states and/or operators as continuous functions of time. Below we will exemplify, that in the continuous limit, $\mathcal{U}[\psi(t)]$ can be identified (essentially) with the factor $\exp\{iS[\psi(t)]/\hbar\}$, where $S[\psi(t)]$ is the action of the system, when propagating along path $|\psi(t)\rangle$. To do so, we introduce fermionic coherent states and start with the simplest case of a non-interacting model, i.e., $\hat{U}_{\delta_t} \equiv :\hat{U}_0(\delta_t):$ by setting $\hat{H}_1 = 0$ in (4.8). The path integral for the full Hamiltonian is derived subsequently, by adding the different transition terms $\hat{H}_{\text{dot}}^U$, $\hat{H}_T$, and $\hat{H}_{\text{int}}^\perp$, one after the other, in that order.

4.1.2. Fermionic Coherent States

Suppose there exists a Hilbert state $|\Psi\rangle$, which is eigenstate of all (electronic) annihilation operators of model (2.1), that is to say, $\hat{d}_\sigma|\Psi\rangle = \mathfrak{d}_\sigma|\Psi\rangle$ and $\hat{c}_{k\sigma p}|\Psi\rangle = \mathfrak{c}_{k\sigma p}|\Psi\rangle$ and analogously for a dual vector $\langle\Psi|$. At the same time, *every* electronic operator $\hat{O}[\hat{d}^\dagger, \hat{d}, \hat{c}^\dagger, \hat{c}]$ can be expressed as a function of the dot and lead electron creators and annihilators. Then, the matrix element of a *normal ordered* operator with two such eigenstates $\langle\Psi'|$ and $|\Psi\rangle$ is given by

$$\langle\Psi'|:\hat{O}[\hat{d}^\dagger, \hat{d}, \hat{c}^\dagger, \hat{c}]:|\Psi\rangle = O(\mathfrak{d}', \mathfrak{d}, \mathfrak{c}', \mathfrak{c})\langle\Psi'|\Psi\rangle, \qquad (4.12)$$

where $O[\mathfrak{d}', \mathfrak{d}, \mathfrak{c}', \mathfrak{c}]$ is a function that is obtained from $\hat{O}$ by replacing all annihilation (creation) operators $\hat{f}$ ($\hat{f}^\dagger$) with the corresponding eigenvalues $\mathfrak{f}$ ($\mathfrak{f}'$). We used $f = c, d$ and introduced $\mathfrak{f} = \mathfrak{c}, \mathfrak{d}$. It is shown below, that states like $|\Psi\rangle$ can be defined properly, but require the introduction of Grassmann numbers and the corresponding algebra, which is reviewed briefly. In the context of constructing a path integral, property (4.12) is then particularly useful, since the matrix element with the short time propagator takes a conveniently simple form. For the simplest case, in which the short time propagator $\hat{U}_{\delta_t}$ equals the normal ordered free propagator $:\hat{U}_0(\delta_t):$, we can write

$$\langle \Psi', \tau' | :\hat{U}_0(\delta_t): | \Psi, \tau \rangle = \exp\left\{-\frac{i}{\hbar} H_0(\mathfrak{d}', \mathfrak{d}, \mathfrak{c}', \mathfrak{c}, \tau) \delta_t\right\} \langle \Psi' | \Psi \rangle \delta_{\tau, \tau'}, \quad (4.13)$$

where $H_0(\mathfrak{d}', \mathfrak{d}, \mathfrak{c}', \mathfrak{c}, \tau)$ is the function obtained from the Hamiltonian by replacing the electron operators with the corresponding eigenvalues [as in (4.12)] and the operator τ_z in $\hat{H}_{\text{imp}}$ be the spin orientation of state $|\tau\rangle$ (interpreted as number). With the states $|\Psi, \tau\rangle$, we introduced the "mixed" basis mentioned above that is used to derive the final version of the path integral.

Consider, for a start, a single fermionic degree of freedom (d.o.f.) with creation and annihilation operators $\hat{f}^\dagger$ and $\hat{f}$, respectively, and anti-commutator relations

$$\{\hat{f}, \hat{f}^\dagger\} = 1 \quad \text{and} \quad \{\hat{f}, \hat{f}\} = 0 = \{\hat{f}^\dagger, \hat{f}^\dagger\}. \quad (4.14)$$

The operator algebra $\mathcal{A}$ generated by $\hat{f}^\dagger$ and $\hat{f}$ is equal to the linear span of $\{\hat{\mathbb{1}}, \hat{f}, \hat{f}^\dagger, \hat{n}\}$ over the field of complex numbers with $\hat{n} = \hat{f}^\dagger \hat{f}$. Any operator $\hat{O}$ that acts on the Hilbert space $\mathfrak{h} = \text{span}(|0\rangle, |1\rangle)$ with $|1\rangle = \hat{f}^\dagger |0\rangle$ can be expressed as linear combination of these four operators. A *fermionic coherent state* $|\mathfrak{f}\rangle$ for a single fermion is defined as an *eigenstate of the annihilation operator* (for more details, see [86, 186], for example):

$$\hat{f}|\mathfrak{f}\rangle = \mathfrak{f}|\mathfrak{f}\rangle \quad \text{and} \quad \langle \mathfrak{f}|\hat{f}^\dagger = \langle \mathfrak{f}|\bar{\mathfrak{f}}, \quad (4.15)$$

where the eigenvalues $\mathfrak{f}$ and $\bar{\mathfrak{f}}$ are (mutually independent) *Grassmann numbers*. Two numbers anti-commute with each other, $\{\mathfrak{f}, \bar{\mathfrak{f}}\} = 0$, as does any number with the fermion operators, $\{\mathfrak{f}, \hat{f}\} = 0$. In particular, $\mathfrak{f}^2 = 0$. Grassmann numbers are generators of a so-called *Grassmann algebra*. For instance, two numbers $\mathfrak{f}$ and $\bar{\mathfrak{f}}$ generate the algebra $\mathfrak{A} = \text{span}\,\mathfrak{b}$, where

$\mathfrak{b} = \{1, \mathfrak{f}, \bar{\mathfrak{f}}, \bar{\mathfrak{f}}\mathfrak{f}\}$. The similarity of $\mathfrak{A}$ and the operator algebra $\mathcal{A}$, suggests to associate[2] their respective generators, that is to say, $\mathfrak{f}$ with $\hat{f}$ and $\bar{\mathfrak{f}}$ with $\hat{f}^\dagger$. In general, for a many-particle Hamiltonian with n fermionic d.o.f., the associated algebra is generated by $2n$ independent Grassmann numbers—one for each creation and annihilation operator. By convention, Grassmann numbers that belong to creation operators are marked with an over-bar. It can be easily checked, that the states

$$|\mathfrak{f}\rangle \equiv (\hat{1} - \mathfrak{f}\hat{f}^\dagger)|0\rangle \quad \text{and} \quad \langle \mathfrak{f}| \equiv \langle 0|(\hat{1} - \hat{f}\bar{\mathfrak{f}}) \quad (4.16)$$

satisfy equation (4.15). And just as every operator, which acts on Hilbert state $\mathfrak{h}$, lies in $\mathcal{A}$, every function of Grassmann numbers $\mathfrak{f}$ and $\bar{\mathfrak{f}}$ is an element of $\mathfrak{A}$ and can thus be expressed as linear combination of elements of $\mathfrak{b}$. It follows, that functions over (subsets of) $\mathbb{R}$ and $\mathbb{C}$ that can be expressed as an infinite series have, when the arguments are Grassmann numbers, always a finite series expansion. As an important example, we illustrate this by means of the exponential function:

$$e^{\mathfrak{f}} = \sum_{n=0}^{\infty} \frac{\mathfrak{f}^n}{n!} = 1 + \mathfrak{f}, \quad e^{\bar{\mathfrak{f}}} = 1 + \bar{\mathfrak{f}}, \quad e^{\bar{\mathfrak{f}} + \mathfrak{f}} = 1 + \bar{\mathfrak{f}} + \mathfrak{f}, \quad \text{and} \quad e^{\bar{\mathfrak{f}}\mathfrak{f}} = 1 + \bar{\mathfrak{f}}\mathfrak{f}.$$
(4.17)

Note that $\exp\{\bar{\mathfrak{f}} + \mathfrak{f}\} \neq \exp\{\bar{\mathfrak{f}}\}\exp\{\mathfrak{f}\} = 1 + \bar{\mathfrak{f}} + \mathfrak{f} + \bar{\mathfrak{f}}\mathfrak{f}$. Since $\langle \mathfrak{f}'|\mathfrak{f}\rangle = 1 + \bar{\mathfrak{f}}'\mathfrak{f} = \exp\{\bar{\mathfrak{f}}'\mathfrak{f}\}$, coherent states are neither normalised nor are states for different d.o.f. even orthogonal to each other.[3] With coherent states the matrix elements of normal ordered fermionic operators $:\hat{O}(\hat{f}, \hat{f}^\dagger): \in \mathcal{A}$

---

[2]Associate, not *identify*. $\mathcal{A}$ and $\mathfrak{A}$ are not isomorphic, due to the differing (anti-)commutation relations.
[3]They all share the unique vacuum state $|0\rangle$.

4. Iterative Summation of Path Integrals

then evaluate to

$$\langle f| :\hat{O}(\hat{f},\hat{f}^\dagger): |f\rangle = (1+\bar{f}f)\,O(\bar{f},f). \tag{4.18}$$

For the over-complete basis of coherent states, a partition of unity can be obtained with the help of properly *defined* notions of the Grassmann differentiation and integration. When acting on the algebra $\text{span}(\{1,f\})$ generated by a single Grassmann number f the derivative operator ∂_f is defined as the *linear operator*

$$\partial_f 1 := 0 \quad \text{and} \quad \partial_f f := 1. \tag{4.19}$$

It resembles the ordinary derivative in the (vector) space of polynomial functions of degree one and less. Derivatives for different generators f and f' inherit their anti-commutative property: $\{\partial_f, \partial_{f'}\} = 0$.

Inspired by the integration over exact differential forms, the Grassmann integration is defined as the linear operator

$$\int df = \int df\, 1 := 0 \quad \text{and} \quad \int df\, f := 1. \tag{4.20}$$

The second identity can be motivated by $\int df\, \partial_f = 0 = \partial_f \int df\, f$. With the requirement that integration is a non-zero operator, the second equality suggests the conclusion: $\int df\, f \propto 1$. Obviously, with definitions (4.19) and (4.20), differentiation and integration are identical operations. These operators are an essential ingredient of

$$\hat{\mathbb{1}} = \partial_{\bar{f}}\,\partial_f\, e^{-\bar{f}f}|f\rangle\langle f| = \int\!\!\int d\bar{f}\,df\, e^{-\bar{f}f}|f\rangle\langle f|, \tag{4.21}$$

the resolution of unity for a single fermionic d.o.f. in the basis of coherent states (see appendix D). The factor $\exp\{-\bar{f}f\}$ cancels the scalar product

4. Iterative Summation of Path Integrals

$\langle f|f\rangle$. For reasons of similarity to conventional resolutions of unity, the integral version is commonly used. This choice also leads to expressions of (4.11) that actually look very similar to ordinary (bosonic) path integrals. Despite this notational compliance, however, the underlying mathematical operations differ largely from each other.

Definition (4.16) can be generalised to many fermionic d.o.f. in a straightforward fashion. For n fermions with annihilators $\hat{f}_i$, $1 \leq i \leq n$ and corresponding creation operators, the coherent states and resolution of unity are given by

$$|\mathfrak{F}\rangle = \prod_{i=1}^{n}(1 - f_i\hat{f}_i^\dagger)|0\rangle \quad \text{and} \quad \langle\mathfrak{F}| = \langle 0|\prod_{i=1}^{n}(1 - \hat{f}_i\bar{f}_i) \quad (4.22a)$$

$$\hat{\mathbb{1}} = \int d\bar{\mathfrak{F}}\, d\mathfrak{F}\, e^{-\bar{\mathfrak{F}}\mathfrak{F}}|\mathfrak{F}\rangle\langle\mathfrak{F}| \equiv \int \prod_{i=1}^{n}\left[d\bar{f}_i\, df_i\, \exp\{-\bar{f}_i f_i\}\right]|\mathfrak{F}\rangle\langle\mathfrak{F}|, \quad (4.22b)$$

where symbol $\mathfrak{F}$, when used like a (Grassmann) number, is the vector $\mathfrak{F} = (f_1, \ldots, f_n)$ and $\bar{\mathfrak{F}}\mathfrak{F} = \sum_i \bar{f}_i f_i$ the scalar product. The symbolic integral measures $d\mathfrak{F} := df_n \cdots df_1$ and $d\bar{\mathfrak{F}} := d\bar{f}_1 \cdots d\bar{f}_n$ are defined in analogy to an Euclidean volume form in the n-dimensional vector space, formed by the vectors $\mathfrak{F}$. The opposite relative ordering with respect to fermion index i ensures, that $\prod_i d\bar{f}_i\, df_i = d\bar{\mathfrak{F}}\, d\mathfrak{F}$. Since creation and annihilation operators appear pairwise in all Hamiltonian parts, so will the Grassmann numbers in the path integral expressions. When this is the case, an analogous Grassmann version of the usual multiplication rule of exponential terms in real and complex space holds (see appendix D):

$$\prod_{i=1}^{n}\exp\{\bar{f}_i f_i\} = \exp\left\{\sum_{i=1}^{n}\bar{f}_i f_i\right\}. \quad (4.23)$$

Based on these remarks, we can define coherent states $\mathfrak{D} := \prod_\sigma(1 -$

$\mathfrak{d}_\sigma \hat{d}^\dagger_\sigma)|0\rangle$ and $\mathfrak{L} := \prod_{k\sigma p}(1-\mathfrak{c}_{k\sigma p}\hat{c}^\dagger_{k\sigma p})|0\rangle$ for the SLQD and the leads, respectively. We then identify the coherent state for the whole electronic system with the tensor product

$$|\Psi\rangle := |\mathfrak{D}\rangle \otimes |\mathfrak{L}\rangle \equiv \prod_{k\sigma p}(1 - \mathfrak{c}_{k\sigma p}\hat{c}^\dagger_{k\sigma p}) \prod_\sigma (1 - \mathfrak{d}_\sigma \hat{d}^\dagger_\sigma)|0\rangle. \quad (4.24)$$

For the "mixed" basis of electron-impurity states, we introduce $|\Psi, \tau\rangle =: |\Psi^\tau\rangle$ as short notation. The unity partition in this basis is given by

$$\hat{\mathbb{1}} = \sum_\tau \int d\bar{\Psi}\, d\Psi\, e^{-\bar{\Psi}\Psi} |\bar{\Psi}^\tau\, \Psi^\tau\rangle. \quad (4.25)$$

4.1.3. The Gaussian Path Integral

Now, by inserting (4.25) on both sides of every short time propagator in (4.9), we arrive at a fermionic path integral representation of the full evolution matrix element. According to (4.13), the resulting expression is particularly simple, when the propagator is normal ordered (which we ensured by definition (4.8)). We start its derivation with the simpler case of the free evolution operator $\hat{U}_0 = \lim_{N\to\infty} :\hat{U}_0(\Delta_t/[N-1]):^{N-1}$. In path integral form, it can be written as (see appendix D)

$$\hat{U}_0(t_f, t_i) = \lim_{\delta_t \to 0} \sum_{\{\tau\}} e^{iS_{\text{imp}}} \int \prod_{j=1}^N [d\bar{\Psi}_j d\Psi_j] \times$$

$$\cdots \times \exp\left\{\sum_{k,l=1}^N \bar{\Psi}_k iG_0^{\text{el}}[\{\tau\}]_{kl}^{-1} \Psi_l\right\} |\Psi_N^\tau\rangle\langle\bar{\Psi}_1^\tau|$$

$$=: \int \mathcal{D}[\bar{\Psi}\Psi\tau] \exp\left\{iS_0[\bar{\Psi}(t), \Psi(t), \tau(t)]/\hbar\right\} |\Psi^\tau(t_f)\rangle\langle\bar{\Psi}^\tau(t_i)|, \quad (4.26)$$

4. Iterative Summation of Path Integrals

where we defined $\{\tau\} := (\tau_N, \ldots, \tau_1)$ and the symbolic measure $\mathcal{D}[\overline{\Psi}\Psi\tau]$ for integration over all continuous paths of the Grassmann vectors $\Psi(t)$, $\overline{\Psi}(t)$ and impurity orientation $\tau(t)$ with $t_i \leq t \leq t_f$. In the continuous notation, the action of the non-interacting model

$$S_0[\overline{\Psi}(t), \Psi(t), \tau(t)]/\hbar = \int_{t_i}^{t_f} \{i\,\overline{\Psi}(t)\partial_t \Psi(t) - H_0[\overline{\Psi}(t), \Psi(t), \tau(t)]/\hbar\}\,\mathrm{d}t \tag{4.27}$$

with $\quad \overline{\Psi}(t)\partial_t \Psi(t) = \sum_\sigma \bar{\mathfrak{d}}_\sigma(t)\partial_t \mathfrak{d}_\sigma(t) + \sum_{k\sigma p} \bar{\mathfrak{c}}_{k\sigma p}(t)\partial_t \mathfrak{c}_{k\sigma p}(t)$

appears. Analogous to (4.13), starting from operator $\hat{H}_0$ the Hamiltonian function $H_0[\overline{\Psi}(t), \Psi(t), \tau(t)]$ at time t is obtained by replacing all fermion operators with corresponding Grassmann fields, e.g., $\hat{d}_\sigma \to \mathfrak{d}_\sigma(t)$, and $\hat{\tau}_z$ with $\tau(t)$. In the exact, discrete version, we separately accounted for the impurity $\hat{H}_{\mathrm{imp}}$ and the rest of the free system involving electronic d.o.f. To this end, we use $\hat{H}_0^{\mathrm{el}} := \hat{H}_0 - \hat{H}_{\mathrm{imp}}$. While S^{imp} is the action of the free impurity, $G_0^{\mathrm{el}}[\{\tau\}]_{jk}^{-1}$ denotes the free inverse Green's matrix with respect to the Hilbert state $|0_\tau\rangle := |0\rangle \otimes |\tau\rangle$ of the "empty" system, i.e., with no electrons in dot and leads. They read

$$iG_0^{\mathrm{el}}[\{\tau\}]_{jk}^{-1} = \begin{pmatrix} -\mathbb{1} & & & & 0 \\ \mathbb{1} - i\Phi_0[\tau_2] & -\mathbb{1} & & & \\ & \mathbb{1} - i\Phi_0[\tau_3] & -\mathbb{1} & & \\ & & \ddots & \ddots & \\ 0 & & & \mathbb{1} - i\Phi_0[\tau_N] & -\mathbb{1} \end{pmatrix} \tag{4.28}$$

and $\quad S^{\mathrm{imp}} = -\dfrac{\Delta_{\mathrm{imp}}\delta t}{2}\sum_{k=2}^{N} \tau_k, \tag{4.29}$

where $\Phi_0[\tau_k] := \mathbf{H}_0^{\text{el}}[\tau_k]\delta_t/\hbar$ and the *diagonal* matrix $\mathbf{H}_0^{\text{el}}[\tau] = E_{0,\varkappa}\delta_{\varkappa,\varkappa'}$, in turn, is implicitly defined by the relation $\overline{\Psi}\mathbf{H}_0^{\text{el}}[\tau]\Psi = H_0^{\text{el}}[\overline{\Psi},\Psi,\tau]$. The index $\varkappa \in \{\sigma\},\{k\sigma p\}$ consecutively numbers the fermion d.o.f. (dot and lead electrons) in the system and $E_{0,\varkappa} = \langle \varkappa | \hat{H}_0^{\text{el}} | \varkappa \rangle$. Apparently, the choice made in definition (4.28), as to which of the consecutive impurity spins τ_k and τ_{k+1} of matrix element $\langle \Psi_{k+1}^\tau | \hat{U}_{\delta_t} | \Psi_k^\tau \rangle$ enter the Φ_0-matrix, is ambiguous. In the expressions shown above, the later spins τ_{k+1} are taken into account. Just as well, we could have chosen the earlier spin or an arithmetic mean of both. As long as a choice is consistent for all matrix elements, this arbitrariness reflects the fact, that the discrete expressions are only accurate up to terms $\mathcal{O}(\delta_t^2)$. Within these bounds, a freedom of choice remains.

For both the action and the Green's matrix, also a discrete and continuous version can be derived, respectively. They are related by

$$S_0[\{\overline{\Psi},\Psi,\tau\}] = \hbar \sum_{k,l=1}^{N} \overline{\Psi}_k G_0^{\text{el}}[\{\tau\}]_{kl}^{-1}\Psi_l + S^{\text{imp}} \quad (4.30)$$

and a corresponding continuous version, which is obtained in the limit $\delta_t \to 0$.[4] A discrete version of the action can be found in appendix D, the (electronic) continuous free inverse Green's function is given by

$$G_0^{\text{el}}[\tau(t)]_{t,t'}^{-1} = \delta(t-t')[i\partial_t - \mathbf{\Omega}_0(t)], \quad (4.31)$$

where $\mathbf{\Omega}_0(t) = \mathbf{H}_0^{\text{el}}[\tau(t)]/\hbar$.

Equation (4.26) represents the free propagator as a *Gaussian path integral*. The name refers to the exponential structure of the integrand, whose argument is a polynomial of second degree in the integration variables, viz., $\overline{\Psi}$ and Ψ. As in the bosonic case, in which the accordant discrete repre-

---
[4] The time step indices are replaced by continuous time parameters and the sums by time integrations.

4. Iterative Summation of Path Integrals

sentation of $\hat{U}_0$ is an integral over complex numbers, fermionic Gaussian integrals are of fundamental importance for the evaluation of physical quantities and, hence, in the course of the following considerations. For an invertible complex $n \times n$-matrix A and a many-particle system with coherent states $|\mathfrak{F}\rangle$ as introduced in (4.22a), the following identity holds (see, e.g., [86, 186]):

$$\int d\bar{\mathfrak{F}}\, d\mathfrak{F}\, \exp\{-\bar{\mathfrak{F}} A \mathfrak{F} + (\bar{\mathfrak{F}} \mathfrak{G} + \bar{\mathfrak{G}} \mathfrak{F})\} = \det(A) \exp\{\bar{\mathfrak{G}} A^{-1} \mathfrak{G}\}, \quad (4.32)$$

where $\bar{\mathfrak{G}}$ and $\mathfrak{G}$ are two arbitrary *independent* Grassmann vectors of dimension n. Note that due to their algebraic structure fermionic path integrals like these always converge. For a singular matrix A, however, this identity cannot be used as the l.h.s. vanishes while the r.h.s. ceases to be a well-defined expression. This is only a seeming contradiction, which appears, since the derivation of the r.h.s. requires, at some point, to factor the term $\det(A)$ out of a polynomial expression (in terms of the matrix elements of A). For singular A, this step would lead to a division by zero (of a polynomial, that is zero as well). But if the Keldysh generating function $\mathcal{Z}[\eta]$ can indeed be expressed in form of a Gaussian integral with regular matrix A, any physical quantity can (at least formally) be evaluated using (4.32).

We illustrate this fact by calculating the matrix element $U_0(f|i) = \langle \psi_f | \hat{U}_0(t_f, t_i) | \psi_i \rangle$ of the free propagator based on equation (4.26). We choose, for simplicity, initial and final state as $|\psi_f\rangle = |\psi_i\rangle = |\mathbf{k}\sigma p, \tau\rangle$, i.e., one electron with wave vector $\mathbf{k}$ and spin σ sits in lead p, the impurity orientation is τ, and the SLQD is empty. Obviously, since $\hat{U}_0(t_f, t_i) = \exp\{-i/\hbar \hat{H}_0 \Delta_t\}$, initial and final state have to be equal to yield a nonvanishing $U_0(f|i)$. It can be evaluated to $\exp\{-i(\epsilon_\mathbf{k} + E_{\text{imp}})\Delta_t/\hbar\}$ with $E_{\text{imp}} = \Delta_{\text{imp}} \tau / 2$. This is the result we want to obtain via Gaussian integration by plugging the discrete version of (4.26) into the definition of $U_0(f|i)$.

4. Iterative Summation of Path Integrals

First of all, it can be seen, that the summation over all tuples $\{\tau\}$ is trivial; since the impurity orientation cannot change in a freely propagating system, only the constant path $(\tau, \ldots, \tau, \tau)$ contributes. Hence, the impurity action evaluates to $-E_{\text{imp}}\Delta_t/\hbar$, which does not depend on δ_t anymore and can be pulled out of the limit. Second, acting with $\langle \psi_f|$ on $|\Psi_N^\tau\rangle$ yields the Grassmann number $c_{\boldsymbol{k}\sigma p}^{(N)}$, the projection of $\langle \Psi_1^\tau |$ on $|\psi_i\rangle$ yields $\bar{c}_{\boldsymbol{k}\sigma p}^{(1)}$. We get

$$
\begin{aligned}
e^{-iS^{\text{imp}}/\hbar} U_0(f|i) &= \langle \boldsymbol{k}\sigma p, \tau | \hat{U}_0 | \boldsymbol{k}\sigma p, \tau \rangle \equiv \langle 0_\tau | \hat{c}_{\boldsymbol{k}\sigma p}(t_f) \hat{c}^\dagger_{\boldsymbol{k}\sigma p}(t_i) | 0_\tau \rangle \\
&= \lim_{\delta_t \to 0} \int \prod_{j=1}^{N} [d\bar{\Psi}_j d\Psi_j] \, c_{\boldsymbol{k}\sigma p}^{(N)} \bar{c}_{\boldsymbol{k}\sigma p}^{(1)} \times \\
&\quad \cdots \times \exp\left\{ \sum_{kl} \bar{\Psi}_k iG_0[\tau]_{kl}^{-1} \Psi_l \right\} \\
&= \lim_{\delta_t \to 0} \frac{\partial}{\partial \eta} \int \prod_{j=1}^{N} [d\bar{\Psi}_j d\Psi_j] \times \\
&\quad \cdots \times \exp\left\{ \sum_{kl} \bar{\Psi}_k iG_0[\tau]_{kl}^{-1} \Psi_l - \eta \bar{c}_{\boldsymbol{k}\sigma p}^{(1)} c_{\boldsymbol{k}\sigma p}^{(N)} \right\}\Bigg|_{\eta=0} \\
&= \lim_{\delta_t \to 0} \frac{\partial}{\partial \eta} \det\{\underbrace{iG_0[\tau]_{kl}^{-1} - \eta \delta_{k,1}\delta_{l,N} \mathbf{X}}_{A_{kl}}\}\Bigg|_{\eta=0},
\end{aligned}
\tag{4.33}
$$

where $\mathbf{X} = \delta_{\varkappa, \boldsymbol{k}\sigma p}\delta_{\varkappa, \varkappa'}$ is implicitly given by equation $\bar{\Psi}_1 \mathbf{X} \Psi_N = \bar{c}_{\boldsymbol{k}\sigma p}^{(1)} c_{\boldsymbol{k}\sigma p}^{(N)}$. In the first line, we identified $U_0(f|i)$ with the (non-interacting) expectation value of the operator product $\hat{c}_{\boldsymbol{k}\sigma p}(t_f) \hat{c}^\dagger_{\boldsymbol{k}\sigma p}(t_i)$ in the empty Hilbert state $|0_\tau\rangle$. It is the probability amplitude to find a single electron in a certain state $|\boldsymbol{k}\sigma p\rangle$ at time t_f, if it was created in the same state at time t_i.

The second line in (4.33) shows the path integral representation of this expectation value, as derived from (4.26). The operators have been replaced by corresponding Grassmann numbers, while the argument of the exponen-

4. Iterative Summation of Path Integrals

tial function is essentially the electronic action S_0^{el} of the considered system (non-interacting and empty). The central idea how to evaluate the integral is shown in the step to line three. Analogous to the procedure shown in section 2.2.2, we first construct a generating function from $\int d\overline{\Psi} d\Psi \exp\{iS_0^{\text{el}}\}$ by adding a source term $i\eta \bar{c}_{k\sigma p}^{(1)} c_{k\sigma p}^{(N)}$ to S_0^{el}. The derivative of this function with respect to η, which is afterwards set to zero (line three), yields back the original expression from line two. Finally, we write the exponent as $\sum_{kl} \overline{\Psi}_k A_{kl} \Psi_l$ and evaluate the Gaussian integral using (4.32) for $\overline{\mathfrak{G}} = \mathfrak{G} = 0$. Matrix A_{kl} is essentially equal to (i times) the Green's matrix (4.28), but with identical $\Phi_0[\tau_k] \equiv \Phi_0$ for all k due to the constant impurity path and the element $-\eta \mathbf{X}$ in the upper right corner (first line, last column). Considering the rather simple structure of $\mathbf{X}$, of the diagonal Φ_0-block, and the whole matrix A_{kl}, it is easy to see, that

$$\begin{aligned}
e^{-iS^{\text{imp}}/\hbar} & U_0(f|i) \\
&= \lim_{\delta_t \to 0} \frac{\partial}{\partial \eta} \det\{A_{kl}\}\Big|_{\eta=0} = \lim_{\delta_t \to 0} \frac{\partial}{\partial \eta} \det\{\mathbb{1} + \eta \mathbf{X}(\mathbb{1} - i\Phi_0)^{N-1}\}\Big|_{\eta=0} \\
&= \lim_{\delta_t \to 0} \frac{\partial}{\partial \eta} \det\left\{\mathbb{1} + \eta \sum_{\varkappa''} \delta_{k\sigma p,\varkappa} \delta_{\varkappa,\varkappa''} \delta_{\varkappa'',\varkappa'} \left(1 - \frac{i}{\hbar} E_{0,\varkappa} \delta_t\right)^{N-1}\right\}\Big|_{\eta=0} \\
&= \lim_{\delta_t \to 0} \left(1 - \frac{i}{\hbar} E_{0,k\sigma p} \delta_t\right)^{N-1} = \exp\left\{-\frac{i}{\hbar} \epsilon_k \Delta_t\right\}.
\end{aligned}$$
(4.34)

In the end, we used $(N-1)\delta_t = \Delta_t$ and $E_{0,k\sigma p} \equiv \epsilon_k$. What is derived here for the free propagator matrix element of a particular Hilbert state, is a basic example of a rather general result: the expectation value of an operator $\hat{O}[\hat{f}^\dagger, \hat{f}]$ in some fermionic system can be represented as path integral over $O[\bar{f}, f]$ times $\exp\{iS\}$ with the action S of the system. The evaluation of this expression with the help of (4.32), however, is only possible, if the action of a general system is quadratic in the Grassmann fields and the corresponding path integral is *Gaussian*. For interacting systems, this condition is in general not met.

In the following subsections, we show how to properly add the three transitions terms (4.5b) to $\hat{H}_0$, to arrive at a Gaussian integral representation of $\hat{U}$ for the full model Hamiltonian (2.1). In this context, we will show, that the main task is to find suitable extensions of the short time propagator. From that, we get a path integral expression similar to (4.26) but for the evolution operator of the interacting system.

4.1.4. Discrete Hubbard-Stratonovich Transformation

First, we deal with the Coulomb term $\hat{H}^U_{\text{dot}}$. Considering it as perturbation $\hat{H}_1$ in equation (4.8) leads to a valid short time propagator and eventually to an exact path integral representation of the evolution operator. Since $\hat{H}^U_{\text{dot}}$ contains a product of *four* fermionic operators instead of two, however, this path integral is not Gaussian and relation (4.32) cannot be used for its evaluation. To fix this, the so-called *discrete Hubbard-Stratonovich transformation* can be used. It constitutes a mapping of the interacting dot electron system to a system of non-interacting electrons, which couple to a virtual field of fluctuating spins. This results in an effective propagator, whose exponent is quadratic in the fermion operators.

Since the commutator $[\hat{H}_0, \hat{H}^U_{\text{dot}}]$ vanishes, the evolution operator for the system with Coulomb interacting dot electrons can be factorised into a free part $\hat{U}_0(\Delta_t)$ and the interacting part $\exp\{-i/\hbar \hat{H}^U_{\text{dot}} \Delta_t\}$. To the latter, we apply the Hubbard-Stratonovich transformation. Details regarding the steps taken are shown in appendix E. Here, we content ourselves with stating the

4. Iterative Summation of Path Integrals

final result:

$$\exp\{-i/\hbar \hat{H}_{\text{dot}}^U \Delta_t\}$$
$$= \frac{1}{2} \sum_{\zeta=\pm 1} \exp\left\{-i/\hbar\left[\frac{U}{2}(\hat{n}_\uparrow + \hat{n}_\downarrow) + i\hbar\zeta\lambda(\Delta_t)(\hat{n}_\uparrow - \hat{n}_\downarrow)\right]\Delta_t\right\} \quad (4.35a)$$

with $\lambda(\Delta_t)\Delta_t = \sinh^{-1}\xi + i\sin^{-1}\xi$, $\xi = \sqrt{\sin[U\Delta_t/(2\hbar)]}$. (4.35b)

Variable ζ can be interpreted as a fluctuating Ising-like spin field and solution (4.35) is unique as long as $0 \leq U\Delta_t/\hbar \leq \pi$. Apparently, this condition limits the usability of (4.35) for large time differences Δ_t. Since we intent to use this identity only to derive a *short time* propagator that leads to a Gaussian integral, this is not an issue. Rather, the uniqueness condition can be seen as imposing an upper bound to the length of the discretization time step δ_t for numerical calculations in dependence on the interaction strength U. The exponential of the r.h.s. in (4.35) commutes with $\hat{H}_0$ and we can write the full propagator as $\exp\{-i/\hbar(\hat{H}_0 + \hat{H}_{\text{dot}}^U)\Delta_t\} = 1/2\sum_\zeta \exp\{-i/\hbar \hat{H}_0^\zeta \Delta_t\}$. The effective free electron (pseudo-) Hamiltonian $\hat{H}_0^\zeta$ results from replacing the dot energies ϵ_σ in $\hat{H}_{\text{dot}}^0$ with

$$\epsilon_\sigma^\zeta(\Delta_t) = \epsilon_\sigma + \frac{U}{2} + i\hbar\sigma\zeta\lambda(\Delta_t) \quad (4.36)$$

in accordance with (4.35a). It is crucial to notice, that due to the presence of an imaginary energy component, $\hat{H}_0^\zeta$ should not be considered as an actual Hamiltonian. From the transformed evolution operator, we get a short time propagator by normal ordering after setting all occurrences of Δ_t to δ_t:[5]

$$\hat{U}_{\delta_t}^U := \frac{1}{2}\sum_{\zeta=\pm 1} :\exp\{-i/\hbar \hat{H}_0^\zeta \delta_t\}: . \quad (4.37)$$

---

[5]Note that the parameter λ depends on the time interval!

4. Iterative Summation of Path Integrals

Using this result, a Gaussian path integral similar to (4.26) can be derived for the Coulomb interacting system. Its path sum then also extends over all tuples $\{\zeta\} := (\zeta_N, \ldots, \zeta_1)$ of the Hubbard-Stratonovich (HS) fields, while the corresponding (pseudo-) Hamiltonian functions contain the modified energies $\epsilon_\sigma^\zeta(\delta_t)$.

4.1.5. Adding the Remaining Interaction Terms

The hybridisation term $\hat{H}_T$ is quadratic in the number of fermions but contains both dot and lead operators. Therefore, the stationary states of the isolated system are in general not eigenvectors of the system with tunnelling. Still, if we set $\hat{H}_1 = \hat{H}_T$ in equation (4.8), the resulting short time propagator is well-formed and leads to the correct Gaussian path integral representation of the full propagator (Coulomb and tunnelling). We could leave it at that. However, since the matrix elements

$$\langle \Psi_k^\tau | : \exp\{-i/\hbar\,\hat{H}_0^\zeta \delta_t\} \left(\hat{\mathbb{1}} - \frac{i}{\hbar}\hat{H}_T \delta_t\right) : | \Psi_j^\tau \rangle$$

and $$\langle \Psi_k^\tau | : \exp\{-i/\hbar\,(\hat{H}_0^\zeta + \hat{H}_T)\delta_t\} : | \Psi_j^\tau \rangle$$

(4.38)

are equal up to second-order in δ_t for all $\langle \Psi_k^\tau |$ and $| \Psi_j^\tau \rangle$, we can as well use the short time propagator that is obtained by adding the coupling term $\hat{H}_T$ to $\hat{H}_0^\zeta$ in definition (4.37). For the further considerations, this variant is chosen for reasons of simplicity.

So far, we had to include the summation over impurity paths $\{\tau\}$ in integrals like (4.26) for formal reasons only. Since no term in $\hat{H}_0^\zeta + \hat{H}_T$ can change the impurity spin orientation, the only path that contributes is constant with all stages equal to the initial value τ_i. With the inclusion of the last remaining transition $\hat{H}_{\text{int}}^\perp$ to our system, this situation changes. This term is responsible for mutual flips of an electron- and the impurity spin, which

4. Iterative Summation of Path Integrals

are also called *flip-flop processes*. It is because of these, that electrons and impurity spin can influence each other non-trivially and dynamical effects such as the relaxation of the impurity spin can appear. As above, we can use equation (4.8) to derive the final version of the short time propagator:

$$\hat{U}_{\delta_t} = \frac{1}{2}\sum_{\zeta} :\exp\{-i/\hbar\,(\hat{H}_0^\zeta + \hat{H}_T)\delta_t\}\left(\hat{\mathbb{1}} - \frac{i}{\hbar}\hat{H}_{\text{int}}^\perp \delta_t\right):. \qquad (4.39)$$

Again, a version written as single exponential $:\exp\{-i/\hbar \hat{H}^\zeta \delta_t\}:$ with $\hat{H}^\zeta := \hat{H}_0^\zeta + \hat{H}_T + \hat{H}_{\text{int}}^\perp$ would be a valid choice, as well. Nevertheless, it is more convenient to use (4.39) when constructing a path integral in the "mixed" basis, since it is separated in parts that contribute to matrix element $\langle \Psi_k^\tau | \hat{U}_{\delta_t} | \Psi_j^\tau \rangle$ either for equal ($\propto \hat{\mathbb{1}}$) or unequal ($\propto \hat{H}_{\text{int}}^\perp$) impurity spin orientations. We want to emphasise, that this is just a calculatory advantage; using the single exponential leads to equivalent results [up to terms $\mathcal{O}(\delta_t^2)$].

In contrast to these variants, operator $\hat{\mathbb{1}} - i/\hbar \hat{H}^\zeta \delta_t$, though fulfilling condition (4.2), is not appropriate to construct a path integral for the model *with flip-flop interaction*.[6] On the level of short time matrix elements, this manifests in the case of opposite left and right impurity spin: the phase factor caused by the free propagation between two instantaneous flip-flop events is missing

$$\langle \Psi_k^\tau | \hat{\mathbb{1}} - i/\hbar \hat{H}^\zeta \delta_t | \Psi_j^\tau \rangle |_{\tau_k \neq \tau_j} = -\frac{iJ\delta_t}{2\hbar}(\delta_{\tau_k,\tau_j+1}\bar{\mathfrak{d}}_\downarrow^k \mathfrak{d}_\uparrow^j + \delta_{\tau_k,\tau_j-1}\bar{\mathfrak{d}}_\uparrow^k \mathfrak{d}_\downarrow^j)e^{\bar{\Psi}_k \Psi_j}. \qquad (4.40)$$

Such an expression would only be correct, if the transition process lasted the whole time span δ_t instead of being instantaneous. In the resulting path integral, the system would not propagate freely between consecutive flip-flop processes in neighbouring time steps and yield an unphysical "continuous

---

[6]It can still be used for the system with Coulomb interaction and tunnelling coupling only.

limit" $\delta_t \to 0$. It is the structure of the "mixed" basis, which leads to this situation and requires a more careful choice of the short time propagator $\hat{U}_{\delta_t}$ than necessary for a pure coherent state basis.

With (4.39), the path integral representation of the full evolution operator $\hat{U}(t_f, t_i)$ can now be derived in the same way as shown in appendix D for the free propagator. It differs from (4.26) mainly by the additional path sum over tuples $\{\zeta\}$ and the additional flip-flop polynomial P of dot electron Grassmann fields in front of the exponential $\exp\{iS/\hbar\}$ with the full action $S[\{\overline{\Psi}, \Psi, \tau, \zeta\}] = \hbar \sum_{kl} \overline{\Psi}_k G^{\text{el}}[\{\tau, \zeta\}]_{kl}^{-1} \Psi_l + S^{\text{imp}}$, where G^{el} is the full electronic Green's matrix for the "empty" system

$$\hat{U}(t_f, t_i) = \lim_{\delta_t \to 0} 2^{-N} \sum_{\{\tau, \zeta\}} \int \prod_{j=1}^{N} [\mathrm{d}\overline{\Psi}_j \mathrm{d}\Psi_j] \, P[\{\tau\}] e^{iS[\{\overline{\Psi}, \Psi, \tau, \zeta\}]/\hbar} |\Psi_N^\tau\rangle\langle\Psi_1^\tau|. \tag{4.41}$$

For a tuple $\{\tau\} = (\tau_N, \ldots, \tau_1)$ with $0 \leq m < N - 1$ flips, the tuple $T_{\text{flip}} := (k_m, \ldots, k_1)$ ascendingly numbers the indices of flip-flop events, where each event $k \in T_{\text{flip}}$ with $\tau_k \neq \tau_{k-1}$ is labelled according to the higher time step index of both spins. The flip-flop polynomial is then given by

$$P[\{\tau\}] := \left(-\frac{iJ\delta_t}{2\hbar}\right)^m \prod_{k \in T_{\text{flip}}} \bar{\mathfrak{d}}_{-\tau_k}^k \mathfrak{d}_{\tau_k}^{k-1}. \tag{4.42}$$

With the help of an examplary impurity path, figure 4.2 illustrates the construction of this polynomial for the whole Keldysh evolution operator (see next section).[7] The inverse Green's matrix $(G^{\text{el}})^{-1}$ has the same structure as shown in equation (4.28). Just the terms $\Phi_0[\tau_k]$ have to be replaced by $\Phi[\tau_k, \zeta_k]$, where in analogy to the free case we implicitly define $\overline{\Psi} \, \Phi[\tau, \zeta] \Psi = (H_0^\zeta + H_T - H_{\text{imp}})[\overline{\Psi}, \Psi, \tau, \zeta] \delta_t/\hbar$. The function $H_{\text{imp}}[\tau] = \Delta_{\text{imp}}\tau/2$ has to be subtracted, as it is already accounted for in the impurity action

---

[7]Equation (4.42) only describes the forward branch.

4. Iterative Summation of Path Integrals

$S_{\text{imp}}.$

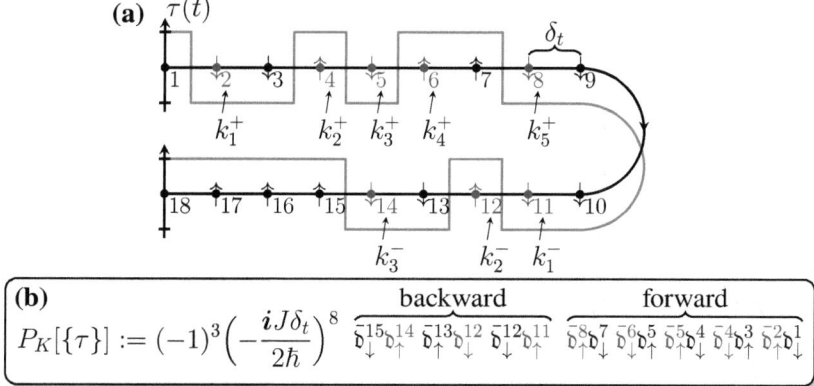

Figure 4.2.: Graphical illustration how to construct the flip-flop polynomials for the examplary impurity path (blue) shown in sub-figure **(a)**. The Keldysh contour (thick line) is divided into $N - 1 = 8$ segments of length δ_t between $2N = 18$ time vertices. The impurity path (tuple of dark and light arrows) features 8 flip-flops with $m^+ = 5$ on the forward (upper) and $m^- = 3$ flips on the backward (lower) branch. As a first step, the flip index tuples $T^\pm_{\text{flip}}$ have to be constructed. Regardless of the contour branch, to each flip-flop the index of the Keldysh time that *is later with respect to the real-time* is assigned. Hence, if two consecutive spins have opposite orientations, the corresponding flip-flop gets the time index of the spin on the right-hand side of the flip (light tint). With this, we have $T^+_{\text{flip}} = (8, 6, 5, 4, 2)$ and $T^-_{\text{flip}} = (14, 12, 11)$. Using equation (4.45), the given impurity path then yields the polynomial shown in sub-figure **(b)**. Note, that compared to the impurity the electrons flips in the opposite direction (not shown).

4.1.6. The Keldysh Partition Function

The last ingredient missing to construct the path integral representation of the Keldysh partition function $\mathcal{Z}$ from the one of the full evolution operator, is a coherent state expression for the trace operation. In appendix D or, e.g.,

4. Iterative Summation of Path Integrals

[186] is shown that for an operator $\hat{O}$ in a fermionic Hilbert space with n d.o.f., we have

$$\text{Tr}\,\hat{O} = \int \mathrm{d}\bar{\mathfrak{F}}\,\mathrm{d}\mathfrak{F}\, e^{-\bar{\mathfrak{F}}\mathfrak{F}} \langle -\mathfrak{F}|\hat{O}|\mathfrak{F}\rangle. \tag{4.43}$$

State $\langle -\mathfrak{F}|$ is obtained from $\langle\mathfrak{F}|$ in (4.22a) by multiplying every Grassmann number $\bar{f}_i$ by -1. Now, the Keldysh partition function $\mathcal{Z}$ can be constructed by plugging (4.41) into (2.11) and use the coherent state version of the trace. In doing so, we have to take into account how the expression for the backwards evolution part of $\hat{U}_K$ has to be modified according to the Hermitian conjugate of the short time propagator (4.39): The Hubbard-Stratonovich (HS) parameter λ is replaced by its complex conjugate and the signs in front of $\Omega[\tau,\zeta]$ in the Green's matrix and in the flip-flop prefactors $\propto J$ have to be inverted. The resulting expression

$$\mathcal{Z}[0] = \overset{(4.43)}{\text{Tr}\{\hat{\rho}(-\infty)} \overset{(4.41)^\dagger}{\hat{U}(-\infty,\infty)} \overset{(4.41)}{\hat{U}(\infty,-\infty)}\}$$

$$= \lim_{\delta_t \to 0} 2^{-2N} \sum_{\{\tau,\zeta\}} \int \prod_{j=1}^{2N} [\mathrm{d}\bar{\Psi}_j \mathrm{d}\Psi_j]\, P_K[\{\tau\}] \exp\{iS_K[\{\bar{\Psi},\Psi,\tau,\zeta\}]/\hbar\}$$

$$\tag{4.44}$$

bears a strong resemblance to the full evolution operator. The polynomial P_K now depends on the Keldysh impurity path $\{\tau\} = (\tau_{2N},\ldots,\tau_1)$. Analogous to the forward evolution operator, $T_{\text{flip}}^+ = (k_{m^+}^+,\ldots,k_1^+)$ is the tuple of ascending flip indices along the forward path $\{\tau^+\} := (\tau_N,\ldots,\tau_1)$ with $\tau_{k^+} \neq \tau_{k^+-1}$ for all $k^+ \in T_{\text{flip}}^+$. Accordingly, $T_{\text{flip}}^- = (k_{m^-}^-,\ldots,k_1^-)$ is the tuple of ascending flip indices along the backward path $\{\tau^-\} := (\tau_{2N},\ldots,\tau_{N+1})$ with $\tau_{k^-} \neq \tau_{k^-+1}$ for all $k^- \in T_{\text{flip}}^-$. Hence, $m = m^+ + m^-$ is the total number of flip-flop events along the Keldysh contour. Note that a flip index on the backward path is labelled according to the *smaller* step index of the flipping spins corresponding to the *later* time. With this, we can write

$$P_K[\{\tau\}] := (-1)^{m^-} \left(-\frac{iJ\delta_t}{2\hbar}\right)^m \prod_{j\in T^-_{\text{flip}}} \bar{\mathfrak{d}}^{j+1}_{\tau_j}\mathfrak{d}^j_{-\tau_j} \prod_{k\in T^+_{\text{flip}}} \bar{\mathfrak{d}}^k_{-\tau_k}\mathfrak{d}^{k-1}_{\tau_k}. \quad (4.45)$$

As before, the discrete Keldysh action is defined with the help of the electronic Green's matrix: it is $S_K[\{\bar{\Psi}, \Psi, \tau, \zeta\}] = S_K^{\text{el}} + S_K^{\text{imp}}$ with $S_K^{\text{el}} = \hbar\sum_{k,l=1}^{2N} \bar{\Psi}_k G_K^{\text{el}}[\{\tau, \zeta\}]_{kl}^{-1} \Psi_l$, where $(G_K^{\text{el}})^{-1}$ relates to the initial density matrix $\hat{\rho}(-\infty)$ and S_K^{imp} is the Keldysh impurity action. They are given by

$$S_K^{\text{imp}} = -\frac{\Delta_{\text{imp}}\delta_t}{2}\sum_{k=2}^{N}(\tau_k - \tau_{2N-k+1}) = -\frac{\Delta_{\text{imp}}}{2}\int_{\mathcal{K}}\tau(t)\,dt \quad (4.46)$$

and

$$iG_K^{\text{el}}[\{\tau,\zeta\}]_{kl}^{-1} =$$

$$\begin{pmatrix}
-\mathbb{1} & & & & & & & -\rho_i \\
\mathbb{1}-i\Phi^+_{[2]} & -\mathbb{1} & & & & & & \\
& \mathbb{1}-i\Phi^+_{[3]} & -\mathbb{1} & & & \langle -\Psi_1|\hat{\rho}(-\infty)|\Psi_{2N}\rangle & & \\
& & \ddots & \ddots & & & & \\
& & & \mathbb{1} & \ddots & & & \\
\langle \Psi_{N+1}|\Psi_N\rangle & & & & & -\mathbb{1} & & \\
& & & & & \mathbb{1}+i\Phi^-_{[3]} & -\mathbb{1} & \\
& & & & & & \mathbb{1}+i\Phi^-_{[2]} & -\mathbb{1}
\end{pmatrix}.$$

(4.47)

We defined the phase terms $\Phi^+_{[k]} := \Phi[\tau_k, \zeta_k]$ and $\Phi^-_{[k]} := \Phi^-[\tau_{2N-k+1}, \zeta_{2N-k+1}]$, where matrix Φ^- is the complex conjugate of Φ. The unity matrix in row $N+1$ and column N is the connection between the forward and backward branch of the Keldysh contour in the infinite future, while the upper right matrix element $-\rho_i$ is the manifestation of initial state $\hat{\rho}(-\infty)$

4. Iterative Summation of Path Integrals

from equation (4.44). The upper left quarter of the Keldysh Green's matrix is identical to the full inverse Green's function that appears in the forward propagator (4.41). The lower right quarter is the corresponding G^{-1} for the backward evolution.

To fix the remaining upper right entry of (4.47), we have to specify the system's initial state. For all further considerations, we assume a configuration of the form

$$\hat{\rho}(-\infty) = |0, \tau_i\rangle\langle 0, \tau_i| \, \hat{\rho}_{\text{leads}}, \tag{4.48}$$

where $\hat{\rho}_{\text{leads}} = \hat{\rho}_L \hat{\rho}_R$ is the equilibrium density matrix of the leads and $|0, \tau_i\rangle$ denotes the empty dot with the impurity in some initial orientation τ_i. For general cases, this is not identical with a thermal equilibrium configuration of the impurity. Rather, it corresponds to a situation, in which the impurity spin has been deliberately prepared in an initial state $|\tau_i\rangle$ (before the dot-lead coupling is switched on) to allow for an easier investigation of the polarisation dynamics. The matrix element of this $\hat{\rho}(-\infty)$ evaluates to (see appendix D)

$$\langle -\Psi_1|\hat{\rho}(-\infty)|\Psi_{2N}\rangle = \mathcal{N}^{-1} \delta_{\tau_1,\tau_i} \delta_{\tau_{2N},\tau_i} \exp\left\{-\sum_{k\sigma p} e^{-\beta(\epsilon_{k\sigma p}-\mu_p)} \bar{c}^1_{k\sigma p} c^{2N}_{k\sigma p}\right\},$$

hence

$$\rho_i = \delta_{\varkappa,\varkappa'} \delta_{\varkappa,k\sigma p} e^{-\beta(\epsilon_{k\sigma p}-\mu_p)}, \tag{4.49}$$

where, as before, the index $\varkappa \in \{\sigma\}, \{k\sigma p\}$ numbers the electrons in the system. The constant $\mathcal{N} = \text{Tr}[\exp\{-\beta \sum_p (\hat{H}_p - \mu_p \hat{N}_p)\}]$ is the normalisation factor of the thermal lead's density matrix (see definition in section 2.3.1) and can be pulled out of the path integral (4.44).

Compared to the case of free propagation (4.26), the introduction and usage of a continuous notation for the full Keldysh partition function requires an even more careful treatment. First of all, a well-defined continuous limit

4. Iterative Summation of Path Integrals

of the discrete paths $\{\tau, \zeta\}$ has to be found.[8] Since the variables τ and ζ can only take the values ± 1, paths $\tau(t)$, $\zeta(t)$ should be elements of the space of non-continuous step functions with $\tau(t) = \pm 1 = \zeta(t)$ and a discrete set of flip points. Second, in the definitions of the continuous action S_K analogous to (4.27)

$$S_K^{\text{el}} = S_{\text{dot}}^{\text{el}} + S_{\text{leads}} + S_T = \hbar \int_{\mathcal{K}} dt dt'\, \bar{\Psi}(t) G_K^{\text{el}}[\tau(t), \zeta(t)]_{t,t'}^{-1} \Psi(t') \quad (4.50\text{a})$$

with

$$S_{\text{dot}}^{\text{el}} = \hbar \sum_\sigma \int_{\mathcal{K}} \bar{\mathfrak{d}}_\sigma(t)[i\partial_t - \Omega_\sigma(t)]\mathfrak{d}_\sigma(t)\, dt \quad (4.50\text{b})$$

$$S_{\text{leads}} = \hbar \sum_{k\sigma p} \int_{\mathcal{K}} \bar{\mathfrak{c}}_{k\sigma p}(t)[i\partial_t - \omega_k]\mathfrak{c}_{k\sigma p}(t)\, dt \quad (4.50\text{c})$$

$$S_T = \sum_{k\sigma p} \int_{\mathcal{K}} [\gamma \bar{\mathfrak{d}}_\sigma(t)\mathfrak{c}_{k\sigma p}(t) + \gamma^* \bar{\mathfrak{c}}_{k\sigma p}(t)\mathfrak{d}_\sigma(t)]\, dt, \quad (4.50\text{d})$$

the fields of the forward and backward parts are unconnected due to the diagonal structure of the continuous inverse Green's function $G_K^{\text{el}}[\tau(t), \zeta(t)]^{-1}$. The discrete version (4.47), on the other hand, connects fields from both contour branches due to the element $\mathbb{1}$ in row $N+1$, column N and—in case of the leads—element $-\rho_i$ in row 1, column $2N$. Thus, in the disconnected lead action (4.50c) any information about the lead's initial state is missing. Kamenev and Levchenko [86] discuss this issue in more detail. Yet, we introduce the continuous notation as it is convenient to use, e.g., for the tracing over the leads (see below). To this end, we define lead and dot frequencies $\omega_k := \epsilon_k/\hbar$ and $\Omega_\sigma(t)$, respectively. The time dependence of the effective dot frequency $\Omega_\sigma(t)$ relates to the impurity and HS field as

---

[8]This is not a problem for the free propagator (4.26), since the impurity path has to be constant anyway.

4. Iterative Summation of Path Integrals

well as to the Keldysh branch: on the forward branch (+), we have

$$\Omega_\sigma(t) = \Omega_\sigma^+(t) \equiv [\epsilon_\sigma + U/2 + J\sigma\tau(t)]/\hbar + i\sigma\zeta(t)\lambda(\delta_t), \qquad (4.51)$$

while on the backward branch (−), we have the complex conjugate $\Omega_\sigma^-(t) = \Omega_\sigma^+(t)^*$.

4.1.7. Adding Source Terms

With the path integral representation of the Keldysh partition function (4.44) and the Green's matrix of the model system (4.47) at hand, we can now construct the generating function $\mathcal{Z}[\eta]$ for some observable $\hat{O}$ that was introduced in section 2.2.2. The method to obtain the expectation value $\langle \hat{O} \rangle(t_{EV})$ from $\mathcal{Z}[\eta]$, which we introduced in equation (2.13), requires to add an appropriate source operator to the exponential of the Keldysh evolution operator in (2.11), followed by a functional derivative.

In the previous sections, we exemplified for the free evolution operator, that this procedure can be applied similarly to a path integral representation. In particular, we want to apply it to expression (4.44). Since we are then dealing only with Grassmann integrals over functions of complex and Grassmann numbers, the resulting expressions are considerably easier to tackle. On the one hand, they do not involve operators but only numbers and, furthermore, Grassmann integrals always converge. Finally, a functional derivative like in equation (2.13) becomes an ordinary derivative with respect to a real number, as in equation (4.34). To construct the generating function for $\langle \hat{O} \rangle(t_{EV})$, we replace every instance of an impurity- or electron operator in observable $\hat{O}[\hat{f}_\varkappa^\dagger, \hat{f}_\varkappa, \hat{\tau}_i]$ by the respective spin- and Grassmann fields from the forward contour branch at the time step closest to t_{EV}. Let us assume that $t_{EV} - t_i = (k-1)\delta_t$. Then, the electron operators are re-

4. Iterative Summation of Path Integrals

placed by their corresponding Grassmann numbers with time step index k. For the Pauli matrices $\hat{\tau}_i$ with $i \in \{x, y, z\}$, the following substitution rules apply

$$\hat{\tau}_i(t_{\text{EV}}) \mapsto \tau_i^{(k)} \quad i \in \{x, y, z\} \tag{4.52}$$

with

$$\tau_x^{(k)} := (1 - \tau_k \tau_{k-1})/2 \quad \tau_y^{(k)} := -i(\tau_k - \tau_{k-1})/2 \quad \tau_z^{(k)} := \tau_k. \tag{4.53}$$

Since matrix elements $\langle \tau' | \hat{\tau}_{x,y} | \tau \rangle$ can only be non-zero for $\tau \neq \tau'$, the Pauli matrices have to be replaced by field expressions that take into account neighbouring spins in time. In other words, only if a flip-flop event occurs at time t_{EV}, the fields $\tau_{x,y}^{(k)}$ can be non-zero. On the forward branch of the Keldysh contour, a flip-flop with $\tau_k = -\tau_{k-1}$ is associated with time step k. Hence, both spins enter the definition of $\tau_{x,y}^{(k)}$. As a consequence, these definitions may seem to be time non-local, for they contain fields that are δ_t apart. Yet, within the accuracy bounds of the path integral expression, they should be considered as quasi-local in time. Just as the Φ_0-matrices in (4.28) "wander" to the diagonal in the continuous version (4.31) of the inverse Green's function, the discrete $\tau_{x,y}^{(k)}$ yield the correct time local continuous limit.

With the resulting $O[\bar{f}_\varkappa^{(k)}, f_\varkappa^{(k)}, \tau_i^{(k)}]$, a generating function $\mathcal{Z}[\eta]$ for $\langle \hat{O} \rangle(t_{EV})$ is obtained from (4.44) by adding $S_O := \eta O$ to the action S_K. The expectation value is evaluated according to

$$\langle O \rangle(t_{\text{EV}}) = -i\partial_\eta \mathcal{Z}[\eta]|_{\eta=0} \equiv -i\partial_\eta \ln \mathcal{Z}[\eta]|_{\eta=0}. \tag{4.54}$$

To increase the conformity with equation (2.13), the factor $-i$ could also

4. Iterative Summation of Path Integrals

be absorbed into the source term. As an example that plays a crucial rule in all further calculations, we present the source term for the charge current I

$$S_I = -\frac{ie\eta}{2} \sum_{k\sigma p} p\big(\gamma \bar{\mathfrak{d}}_\sigma^{(k)} \mathfrak{c}_{k\sigma p}^{(k)} - \gamma^* \bar{\mathfrak{c}}_{k\sigma p}^{(k)} \mathfrak{d}_\sigma^{(k)}\big). \tag{4.55}$$

Its explicit form has to be considered at this point, since it has to be added before the lead d.o.f. are traced out in the next section. This is actually the case for all observables that contain lead electron operators. However, the only example of such an observable considered in this work is the charge current.

4.1.8. Tracing over the Electron Degrees of Freedom

As we explained in chapter 2, it is one of the crucial model features, that the leads stay in equilibrium at all times, and the systems dynamics only appears in the quantum dot part. We already made use of this fact in section 2.3.1 by tracing over all lead degrees of freedom to arrive at effective rates for transition processes between dot states. In doing so, the model with infinitely many d.o.f. was mapped to the small SLQD subsystem. After choosing the initial state (4.49), we can do the same for the generating function $\mathcal{Z}[\eta]$ within the path integral framework. To this end, we perform the path integral over all Grassmann numbers $\bar{\mathfrak{c}}_{k\sigma p}, \mathfrak{c}_{k\sigma p}$ associated with lead electrons. This can be done with the help of identity (4.32). According to equation (4.50), the full action $S_K^{\text{el}} + S_I$ is the sum of terms containing either dot fields only ($S_{\text{dot}}^{\text{el}}$), lead fields only ($S_{\text{leads}}$), or a mixture of both ($S_T + S_I$). For convenience, we cut the Keldysh contour "in half" considering the forward and backward branch separately. Then, the resulting expressions are integrals over the real time contour from $-\infty$ to ∞. In doing so, each time dependent field acquires a branch index $\pm$ indicating whether it belongs to

4. Iterative Summation of Path Integrals

the forward (+) or backward branch (−). Accordingly, expressions like the Green's matrices below gain an additional 2 × 2-superstructure:

$$
\begin{aligned}
S_K^{\text{el}} = \hbar \int_{-\infty}^{\infty} dt \int_{-\infty}^{\infty} dt' \, \{ & \bar{\mathfrak{D}}(t)(G_{\text{dot}}^{\text{el}})_{t,t'}^{-1} \mathfrak{D}(t') + \bar{\mathfrak{L}}(t)(G_{\text{leads}})_{t,t'}^{-1} \mathfrak{L}(t') \\
& + \bar{\mathfrak{D}}(t) F_{t,t'}^{\dagger} \mathfrak{L}(t') + \bar{\mathfrak{L}}(t) F_{t,t'} \mathfrak{D}(t') \}
\end{aligned}
\quad (4.56a)
$$

with

$$
(G_{\text{dot}}^{\text{el}})_{t,t'}^{-1} = \delta(t-t') \mathbb{1}_{\text{dot}} \begin{pmatrix} i\partial_t - \Omega_\sigma^+(t) & 0 \\ 0 & -i\partial_t + \Omega_\sigma^-(t) \end{pmatrix} \quad (4.56b)
$$

$$
(G_{\text{leads}})_{t,t'}^{-1} = \delta(t-t') \mathbb{1}_{\text{leads}} \begin{pmatrix} i\partial_t - \omega_k & 0 \\ 0 & -i\partial_t + \omega_k \end{pmatrix} \quad (4.56c)
$$

$$
F_{t,t'} = \frac{\gamma^*}{\hbar} \left[1 + \frac{ie\eta p}{2} \delta(t - t_{\text{EV}}) \right] \delta(t-t') \delta_{\varkappa,k\sigma p} \delta_{\varkappa',\sigma'} \delta_{\sigma,\sigma'} \begin{pmatrix} 1 & 0 \\ 0 & -1 \end{pmatrix}. \quad (4.56d)
$$

The definition of the matrix $F_{t,t'}$ is chosen in such a way, that the integral over the last two terms of equation (4.56a) yields $S_T + S_I$, i.e., the part of the action mixing both dot and lead electron fields. The minus signs of the (− −) components in each matrix account for the (originally) negative direction of integration on the backward Keldysh branch. As the Green's functions gain a matrix superstructure, the fields $\bar{\mathfrak{F}} = (\bar{\mathfrak{F}}^+, \bar{\mathfrak{F}}^-)$ and $\mathfrak{F} = (\mathfrak{F}^+, \mathfrak{F}^-)^T$ with $\mathfrak{F} = \mathfrak{L}, \mathfrak{D}$ become 2-component vectors, when integrated over the real time contour. If we further define

$$
\mathfrak{J}(t) := \int_{-\infty}^{\infty} dt' F_{t,t'} \mathfrak{D}(t') = \frac{\gamma^*}{\hbar} \left[1 + \frac{ie\eta p}{2} \delta(t - t_{\text{EV}}) \right] \delta_{\varkappa,k\sigma p} \begin{pmatrix} \mathfrak{d}_\sigma^+ \\ -\mathfrak{d}_\sigma^- \end{pmatrix} (t) \quad (4.57)
$$

and an analogous vector $\bar{\mathfrak{J}}$, the part of path integral (4.44) that involves lead fields can be written in a form manifestly equivalent to the l.h.s. of equation

4. Iterative Summation of Path Integrals

(4.32). With the use of (4.56) and (4.57), we can identify $A = -iG_{\text{leads}}^{-1}$, the fields $\mathfrak{F} = \mathfrak{L}$, and the "arbitrary" vectors $\mathfrak{G}$ and $\bar{\mathfrak{G}}$ with $\mathfrak{J}$ and $\bar{\mathfrak{J}}$, respectively. Hence, integration over the lead degrees of freedom yields

$$\int \mathcal{D}[\bar{\mathfrak{L}}\mathfrak{L}] e^{i(S_{\text{leads}} + S_T + S_I)/\hbar}$$
$$= \det\{-iG_{\text{leads}}^{-1}\} \exp\{i \overbrace{\int_{-\infty}^{\infty} dt \int_{-\infty}^{\infty} dt' \, \bar{\mathfrak{J}}(t) (G_{\text{leads}})_{t,t'} \mathfrak{J}(t')}^{S_{\text{env}}/\hbar}\}. \quad (4.58)$$

To further evaluate the r.h.s. of this equation, we have to calculate the lead's Green's function and the determinant of its inverse. This requires using the discrete version of the Green's matrix, because inversion of the diagonal expression (4.56c) results in vanishing $(+-)$ and $(-+)$ components. And what is more, the diagonal components obtained this way contain no information about the initial state. As we lay out in appendix F, however, the Keldysh components of the Green's function are given by

$$\begin{aligned}
(G_{\text{leads}}^{+-})_{t,t'}^{R,R'} &= ie^{-i\epsilon_k(t-t')/\hbar} f_p^+(\epsilon_k) \delta_{R,R'} \\
(G_{\text{leads}}^{-+})_{t,t'}^{R,R'} &= -ie^{-i\epsilon_k(t-t')/\hbar} f_p^-(\epsilon_k) \delta_{R,R'} \\
(G_{\text{leads}}^{++})_{t,t'}^{R,R'} &= \theta(t-t') (G_{\text{leads}}^{-+})_{t,t'}^{R,R'} + \theta(t'-t) (G_{\text{leads}}^{+-})_{t,t'}^{R,R'} - i\delta_{t,t'}/2 \\
(G_{\text{leads}}^{--})_{t,t'}^{R,R'} &= \theta(t-t') (G_{\text{leads}}^{+-})_{t,t'}^{R,R'} + \theta(t'-t) (G_{\text{leads}}^{-+})_{t,t'}^{R,R'} - i\delta_{t,t'}/2,
\end{aligned}$$
(4.59)

where $f_p^{\pm}$ are the leads Fermi functions introduced in section 2.3.1. We introduced $\boldsymbol{R} := (k\sigma p)$ for short and the step-function $\theta(t)$ with $\theta(0) := 1/2$. Note that $\delta_{t,t'}$ denotes the *Kronecker delta* rather than the delta distribution. The Green's function can also be calculated directly with definition $(G_{\text{leads}}^{\alpha\beta})_{t,t'} := i\langle \hat{T}_K \hat{c}_R^{\alpha}(t) \hat{c}_R^{\dagger,\beta}(t') \rangle$ using the Keldysh time ordering operator $\hat{T}_K$; the $\alpha, \beta = \pm$ are contour indices. Also in appendix F is shown, that

4. Iterative Summation of Path Integrals

the determinant of the inverse Green's matrix evaluates to

$$\det\{-iG^{-1}_{\text{leads}}\} = \prod_{k\sigma p}\left(1 + e^{-\beta(\epsilon_k - \mu_p)}\right) \equiv \mathcal{N} \qquad (4.60)$$

and therefore cancels the normalisation constant in (4.49) that was introduced with the thermal initial state of the leads. An important relation that is connected to causality and can be read out of (4.59) is $G^{++}_{\text{leads}} + G^{--}_{\text{leads}} - G^{+-}_{\text{leads}} - G^{-+}_{\text{leads}} = -i\delta_{t,t'}$.

After the integration, the exponent on the r.h.s. of (4.58) contains dot electron fields only and has the form of an action S_{env}. It can be interpreted as comprising the effects of the thermal leads (as environment) on the dot and takes the form

$$S_{\text{env}} = \hbar \sum_{\sigma p} \int_{-\infty}^{\infty} dt \int_{-\infty}^{\infty} dt'\, \bar{\mathfrak{d}}_\sigma(t)\boldsymbol{\gamma}(p, t-t')\left\{1 + \frac{ie\eta p}{2}[\delta_{\text{EV}}(t') - \delta_{\text{EV}}(t)]\right\}\mathfrak{d}_\sigma(t') \qquad (4.61a)$$

with the Keldysh matrix

$$\boldsymbol{\gamma}(p, t-t') = \frac{|\gamma|^2}{\hbar^2} \sum_k \begin{pmatrix} G^{++}_{\text{leads}} & -G^{+-}_{\text{leads}} \\ -G^{-+}_{\text{leads}} & G^{--}_{\text{leads}} \end{pmatrix} (kp; t-t') \qquad (4.61b)$$

and the short notation $\delta_{\text{EV}}(t) := \delta(t - t_{\text{EV}})$. It is important to keep in mind, that t_{EV} is fixed on the forward contour branch. Therefore the delta distributions $\delta_{\text{EV}}(t)$ and $\delta_{\text{EV}}(t')$ only appear in the first row and column, respectively. Since η has to be considered infinitesimal as far as the evaluation of observables is considered, the term proportional to $\eta^2 \delta_{\text{EV}}(t)\delta_{\text{EV}}(t')$ in (4.61) can be neglected. The minus signs of the $(+-)$ and $(-+)$ components of the matrix $\boldsymbol{\gamma}$ originate again from the inverse integration direction along the backward branch. We perform the wide-band limit substitution $\sum_k \to \varrho(\epsilon_F)\int d\epsilon_k$ to evaluate it further and, for $t - t' \neq 0$, arrive at (see appendix F)

4. Iterative Summation of Path Integrals

$$\gamma(p, t - t') = \frac{\Gamma}{2\beta\hbar^2} \frac{e^{-i\mu_p(t-t')/\hbar}}{\sinh[\pi(t - t')/(\hbar\beta)]} \begin{pmatrix} -1 & 1 \\ 1 & -1 \end{pmatrix}. \tag{4.62}$$

We have to restrict the use of this equation to non-vanishing time differences, since the real-time γ-matrix shows a $(t - t')^{-1}$ divergence. This turns out to be a problem for the numerical calculations, since the large matrix elements for small $(t - t')$ can cause numerical instabilities. In section 4.2.3, we show how this problem can be avoided. For large time differences on the other hand, the absolute value of γ decays exponentially. As can be seen in section 4.2.1, this important property can be exploited in a numerical scheme and is the cornerstone of the ISPI method.

After the integration over the lead fields, the generating function $\mathcal{Z}[\eta]$ for the charge current only depends on the dot fields $\overline{\mathfrak{D}}(t)$ and $\mathfrak{D}(t)$. The next step is to integrate over the remaining dot fields, to arrive at a formally exact result for the generating function. To do so, we have to write the environmental action with some (effective) environmental inverse Green's function G_{env}^{-1} in order to perform the Gaussian integration according to (4.32). By plugging (4.62) into (4.61), we obtain

with
$$S_{\text{env}} = \hbar \int_{-\infty}^{\infty} dt \int_{-\infty}^{\infty} dt' \, \overline{\mathfrak{D}}(t) \left(G_{\text{env}}\right)_{t,t'}^{-1} \mathfrak{D}(t')$$
$$\left(G_{\text{env}}\right)_{t,t'}^{-1} = \left(G_{\text{env}}^0\right)_{t,t'}^{-1} + \eta \left(G_{\text{env}}^I\right)_{t,t'}^{-1} \tag{4.63a}$$

and

$$\left(G_{\text{env}}^I\right)_{t,t'}^{-1} =$$
$$\mathbb{1}_{\text{dot}} \frac{e\Gamma}{2\beta\hbar^2} \frac{\sin[eV_{\text{bias}}(t - t')/(2\hbar)]}{\sinh[\pi(t - t')/(\hbar\beta)]} \begin{pmatrix} \delta_{\text{EV}}(t) - \delta_{\text{EV}}(t') & -\delta_{\text{EV}}(t) \\ \delta_{\text{EV}}(t') & 0 \end{pmatrix}, \tag{4.63b}$$

where we used $\mu_p = peV_{\text{bias}}/2$. Despite the singularity at $t - t' = 0$ of

4. Iterative Summation of Path Integrals

the individual matrix $\gamma(p, t - t')$, their difference for $p =$ L and R, as it appears in the part of S_{env} that is proportional to η, is finite for all values of t and t'. The matrix $(G_{\text{env}}^0)^{-1}$, however, is divergent for $t = t'$. It is for this reason, that we will not explicitly use its real-time form throughout the following derivations and refrain from presenting it here.[9] At this point, source terms for observables $\hat{O}$ that only contain dot fields can be added to the action. This is equivalent to adding some appropriate $\eta(G^O)^{-1}$ to the inverse Green's function. Depending on which observable is of interest, the effective full inverse Green's function $(G^{\text{eff}})^{-1}$ is then given either by $(G_{\text{dot}}^{\text{el}})^{-1} + (G_{\text{env}})^{-1}$ for the current or $(G_{\text{dot}}^{\text{el}})^{-1} + \eta(G^O)^{-1} + (G_{\text{env}}^{\text{el}})^{-1}$ otherwise. With this, we can write the remaining path integral as

$$\int \mathcal{D}[\bar{\mathfrak{D}}\mathfrak{D}] \, P_K \exp\Big\{ \overbrace{i \int_{-\infty}^{\infty} dt \int_{-\infty}^{\infty} dt' \, \bar{\mathfrak{D}}(t) \big(G^{\text{eff}}\big)^{-1}_{t,t'} \mathfrak{D}(t')}^{S_K^{\text{eff}}/\hbar} \Big\}$$

$$= \langle P_K \rangle \det\{(iG^{\text{eff}})^{-1}\}. \qquad (4.64)$$

This equation is actually the general definition of the expectation value of a polynomial P_K of Grassmann numbers in a system with Green's function G^{eff}. It is similar to equations (4.33) and (4.54) but in contrast to those, here we have to explicitly account for the normalising determinant $\det\{(iG^{\text{eff}})^{-1}\}$. This is because the impurity and HS paths, which are involved in an individual contribution to $\mathcal{Z}$, are in general not symmetric on the forward and backward branch. As a consequence, the determinant of the effective Green's matrix is unknown, while the respective constants are 1 both for the free propagator matrix element $(\det iG_0^{-1})$ and the full Keldysh expectation value $(\mathcal{Z}[0])$.

So far, we did nothing more than just rewriting the path integral. The next step is to derive an expression for $\langle P_K \rangle$ that refers solely to G^{eff}. This

---

[9] It is essentially given by the sum $\gamma(\text{L}) + \gamma(\text{R})$ and differs from (4.63) by having a cosine instead of a sine in the numerator.

4. Iterative Summation of Path Integrals

can be done with the help of *Wick's theorem*. A version that is based on the path integral can be found in [86]. First, we exploit the fact that, when an expectation value is considered, fermion operators and Grassmann fields are interchangeable. This property is "built in" by construction of the path integral and can be seen in equation (4.33). With the use of (4.45), we get

$$\langle P_K[\tau(t)]\rangle \propto \langle \prod_{j\in T^-_{\text{flip}}} \hat{d}^\dagger_{\tau_j}(t_j^<)\hat{d}_{-\tau_j}(t_j) \prod_{k\in T^+_{\text{flip}}} \hat{d}^\dagger_{-\tau_k}(t_k)\hat{d}_{\tau_k}(t_k^<)\rangle, \qquad (4.65)$$

where $\tau_k = \tau(t_k + 0^+)$ is the value of the impurity field at the flip time t_k. By this definition, we fix the value of τ immediately at the real-time later time step. The operators are ordered with ascending *Keldysh times* from right to left, i.e., with larger index number their *real-times* decrease on the backward branch. The superscript "$<$" indicates, that the respective time lies an infinitesimal real-time step in the past: $t_k^< = t_k - \mathrm{d}t$. This is the continuous limit $\delta_t \to \mathrm{d}t$ of the discrete expression with flip times lying δ_t apart. Though it can be neglected in the continuous version, this information is used to properly re-discretise the following result for the subsequent numerical calculations. The first important insight that we gain from the r.h.s. of (4.65) is the fact, that $\langle P_K[\tau(t)]\rangle$ vanishes for an odd number m of flip-flops. This is because every flip-flop contributes both a creator and an annihilator for electrons with opposite spins to P_K. Therefore, a chain of m consecutive flip-flops yields an equally long alternating product of creators and annihilators for each spin σ. If m is odd, so is the length of these alternating products, which therefore contain an unequal number of $\hat{d}^\dagger$ and $\hat{d}$. When applied to any state $|\psi\rangle$ in the trace (that is the expectation value), such an operator product changes the particle count by 1, so that the projection with $\langle\psi|$ always vanishes. Thus, we only have to consider paths with even m, from now on.

4. Iterative Summation of Path Integrals

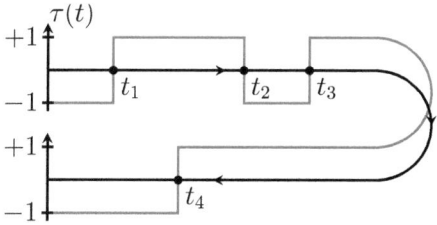

Figure 4.3.: Examplary impurity path (gray) with 4 flip-flop events along the Keldysh contour (black). Note, that contrary to the time-descrete path in figure 4.2, the Keldysh contour is continuous and the (equally continuous) flip-flop times $t_1, \ldots, t_4$ should not be confused with the discrete time vertices.

We start by illustrating how an expression like (4.65) can be evaluated using Wick's theorem with the impurity path shown in figure 4.3. Afterwards, the expression for arbitrary $\tau(t)$ is derived. The exemplary path features a total of four flip-flops along the contour, three on the forward, one on the backward branch. Thus, up to prefactors, the flip-flop polynomial becomes

$$\begin{aligned}\langle P_K \rangle &\propto \langle \hat{d}_\uparrow^\dagger(t_4^<)\hat{d}_\downarrow(t_4)\hat{d}_\downarrow^\dagger(t_3)\hat{d}_\uparrow(t_3^<)\hat{d}_\uparrow^\dagger(t_2)\hat{d}_\downarrow(t_2^<)\hat{d}_\downarrow^\dagger(t_1)\hat{d}_\uparrow(t_1^<)\rangle \\ &= \underbrace{\langle \hat{d}_\uparrow^\dagger(t_4^<)\hat{d}_\uparrow(t_3^<)\hat{d}_\uparrow^\dagger(t_2)\hat{d}_\uparrow(t_1^<)\rangle}_{P_K^\uparrow} \underbrace{\langle \hat{d}_\downarrow(t_4)\hat{d}_\downarrow^\dagger(t_3)\hat{d}_\downarrow(t_2^<)\hat{d}_\downarrow^\dagger(t_1)\rangle}_{P_K^\downarrow}. \end{aligned} \quad (4.66)$$

Since the action S_K^{eff} can be decomposed into disparate parts S_σ^{eff} with associated Green's matrices G_σ^{eff} for both electron spins σ, we are allowed to factorise the expectation value of the mixed operator product accordingly. This is possible even if a source action S^O contains terms mixing fields of different spins. Since we do not study any such observables in this work, we do not elaborate on this point in detail but content ourselves with a qualitative explanation as to why this is. An observable $\hat{O}$ that mixes both spins

4. Iterative Summation of Path Integrals

is either quadratic in the dot electron operators or proportional to $\hat{n}_\uparrow \hat{n}_\downarrow$.[10] If it is quadratic, we can construct the path integral directly from equation (2.8) in section 2.2.2. That would result in an expression that is equal to $\mathcal{Z}[0]$ up to an additional factor $O[\bar{\mathfrak{d}}, \mathfrak{d}, \tau](t_{\text{EV}})$, which can be absorbed into P_K. An observable $\hat{O}$ proportional to $\hat{n}_\uparrow \hat{n}_\downarrow$ can be dealt with by absorbing it into the Coulomb Hamiltonian prior to the Hubbard-Stratonovich transformation. Applying Wick's theorem to (4.66) then yields (see appendix F)

$$\langle P_K^\uparrow \rangle = -\det \underbrace{\begin{pmatrix} (G_\uparrow^{\text{eff}})_{t_1^<,t_2}^{++} & (G_\uparrow^{\text{eff}})_{t_1^<,t_4^<}^{+-} \\ (G_\uparrow^{\text{eff}})_{t_3^<,t_2}^{++} & (G_\uparrow^{\text{eff}})_{t_3^<,t_4^<}^{+-} \end{pmatrix}}_{\Xi_\uparrow}$$

and $$\langle P_K^\downarrow \rangle = -\det \underbrace{\begin{pmatrix} (G_\downarrow^{\text{eff}})_{t_2^<,t_1}^{++} & (G_\downarrow^{\text{eff}})_{t_2^<,t_3}^{++} \\ (G_\downarrow^{\text{eff}})_{t_4,t_1}^{-+} & (G_\downarrow^{\text{eff}})_{t_4,t_3}^{-+} \end{pmatrix}}_{\Xi_\downarrow}.$$

(4.67)

From these expressions, we can extract the information needed to derive a general result for a path $\tau(t)$ with flip-flops at times $T_{\text{flip}} = (t_m, \ldots, t_1)$ with m even. Note, that in continuous notation the entries of T_{flip} are actual (continuous) flip-flop times (more precisely, they are *real-times* with an index reflecting their *Keldysh order*) and not time indices as in section 4.1.5. It is only in the course of the subsequently explained re-discretisation of these expressions, that the flip times are mapped to time-step indices.

Given the initial impurity orientation $\tau_i = \tau(-\infty)$, the matrix Ξ_{τ_i} of dimension $m/2$ is constructed by first tagging every other element of T_{flip}, but separately for each contour branch, with an "<" superscript. In doing so, we have to ensure that (i) the second flip on the forward contour receives a "<" and (ii) the last time on the forward and the first time (in Keldysh

---
[10] Due to the small Hilbert space and to particle number conservation

4. Iterative Summation of Path Integrals

order) on the backward contour have the same superscript. With ascending step indices, the times of even *numbered* flips are assigned to the rows and the times of odd numbered flips to the columns. The entries of the matrix are then filled with the elements of the Green's matrix $G^{\text{eff}}_{\tau_i}$ at the respective times and with the corresponding contour indices. Accordingly, to build matrix $\Xi_{-\tau_i}$, times of odd numbered flip times are assigned to the rows, the even ones to the columns and all times are, compared to Ξ_{τ_i}, reciprocally tagged with "<". This is shown in figure 4.4(a) and (c) for the Ξ_{τ_i}-matrix and an example path.

At this point, it is useful to re-discretise the result, which is a prerequisite for the numerical treatment. To this end, we have to replace the flip-flop times t_j with time step indices $k_j = k_j^+$ and $k_j = k^-_{j-m^+}$ on the forward ($j \leq m^+$) and backward ($j > m^+$) branch, respectively. They were introduced in section 4.1.6. For the "lesser" times $t_j^<$, we have to substitute $k_j^< := k_j - \alpha$, where $\alpha = \pm 1$ is the branch index of time step k_j. After these steps, we can write down the final expression for the generating function as

$$\mathcal{Z}[\eta] = \lim_{\delta_t \to 0} 2^{-2N} \sum_{\{\tau,\zeta\}} (-1)^{m^-} \left(\frac{J\delta_t}{2\hbar}\right)^m e^{i(S_K)_{\text{imp}}} \prod_\sigma \det i(G^{\text{eff}}_\sigma)^{-1} \det \Xi_\sigma$$

with $\quad (\Xi_{\tau_i})_{qr} = (G^{\text{eff}}_{\tau_i})_{\overline{K}^{\text{E}}_q K^{\text{O}}_r}, \quad$ and $\quad (\Xi_{-\tau_i})_{qr} = (G^{\text{eff}}_{-\tau_i})_{\overline{K}^{\text{O}}_q K^{\text{E}}_r}.$

(4.68)

The summation over impurity paths is restricted to tuples $\{\tau\}$ with $\tau_1 = \tau_{2N} = \tau_i$. We defined the m dimensional tuples $\boldsymbol{K} := (k_m^<, \ldots, k_{m^++1}^<, k_{m^+}, \ldots, k_1)$ and $\overline{\boldsymbol{K}} := (k_m, \ldots, k_{m^++1}, k_{m^+}^<, \ldots, k_1^<)$. The corresponding vectors with superscript E (O) denote the $m/2$ dimensional sub-vectors consisting of only the even (odd) indices. Note that, since they refer to times in Keldysh space, i.e., $1 \leq k_j \leq 2N$ for all j, the Green's matrix elements in the definition of the Ξ matrices have no superscript branch indices. In general, the elements of a Green's matrix in Keldysh- and real-time space

4. Iterative Summation of Path Integrals

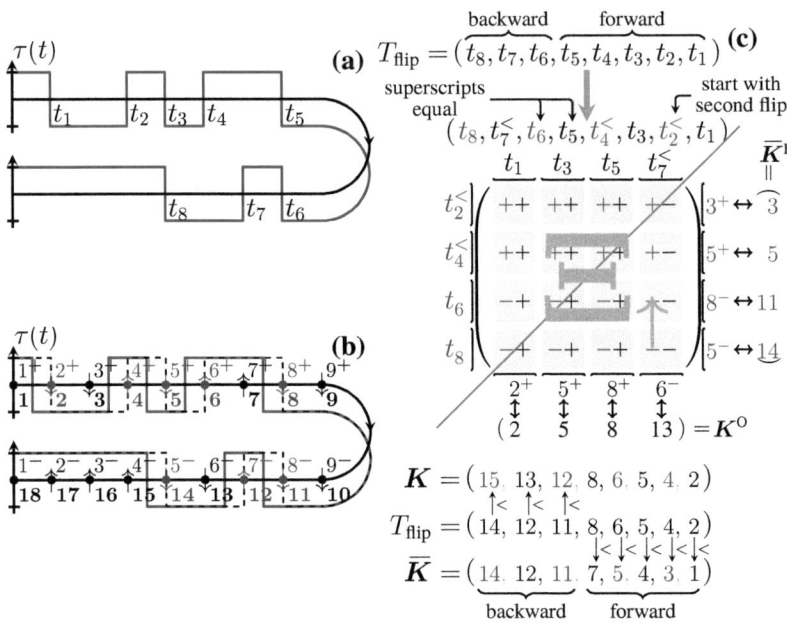

Figure 4.4.: Formal construction rules for the Ξ-matrices. Both for a continuous impurity path [blue in **(a)**] and the discrete version with $N = 9$ [sub-figure **(b)**, identical to figure 4.2], we determine the q, r and α, α' of the matrix elements $(G^{\text{eff}}_{\tau_i})^{\alpha\alpha'}_{qr}$ of Ξ_{τ_i}, where $\tau_i = \uparrow$. In **(a)**, t_1 to t_8 are the *real-times* of the 8 flip-flops, numbered according to their *Keldysh order*. The flips of the discrete version are shifted to the left by $\delta_t/2$ compared to the continuous path, as they have to lie *between two neighbouring* vertices. The upper (left) part of sub-figure **(c)** shows how to proceed in the continuous case. First, every other time of T_{flip} is tagged with superscript '<', starting from the second flip time t_2, where the last time on the forward- and first on the backward branch must have equal tags (none, in this case). Of the resulting tuple of real-times, the even numbered entries (gray) are assigned to the rows, the odd numbered ones to the columns of $\Xi_\uparrow$. Their contour indices determine α and α' (in the shaded squares). For the discrete path, arrows marked with '<' indicate how K and $\overline{K}$ are derived from the tuple of step indices T_{flip} in the lower (right) part of **(c)**.

138

4. Iterative Summation of Path Integrals

are related by

$$G_{qr} = G_{\tilde{q}\tilde{r}}^{\alpha(q)\alpha(r)} \begin{cases} \alpha(q) = + \text{ and } \tilde{q} = q & \text{for } 1 \leq q \leq N \\ \alpha(q) = - \text{ and } \tilde{q} = 2N - q + 1 & \text{for } N < q \leq 2N \end{cases}. \tag{4.69}$$

In figure 4.4(b) and (c), we show how to properly assign the step- and contour indices to the rows and columns of the elements of $\Xi_\uparrow$ for the same discrete impurity path as shown in figure 4.2. Obviously, the row and column indices correspond to the $\mathfrak{d}_\uparrow^q$ and $\bar{\mathfrak{d}}_\uparrow^r$ fields (with ascending q and r), respectively, of the flip-flop polynomial $P_K[\{\tau\}]$ in figure 4.2(b). Hence, those formal rules just implement the following procedure to assign the proper time-step (and contour-) indices to the rows and columns of Ξ_τ. For a given impurity path $\{\tau\}$ with even m and $\tau_1 = \tau_{2N} = \tau_i$:

(4.R1) Construct the flip-flop polynomial $P_K[\{\tau\}]$

(4.R2) Assign step indices $q_1 < \ldots < q_{m/2}$ of (annihilator) fields $\mathfrak{d}_\tau^{q_1}, \mathfrak{d}_\tau^{q_2}, \ldots, \mathfrak{d}_\tau^{q_{m/2}}$ that appear in $P_K[\{\tau\}]$ to the *rows* of Ξ_τ.

(4.R3) Assign step indices $r_1 < \ldots < r_{m/2}$ of (creator) fields $\bar{\mathfrak{d}}_\tau^{r_1}, \ldots, \bar{\mathfrak{d}}_\tau^{r_{m/2}}$ that appear in $P_K[\{\tau\}]$ to the *columns* of Ξ_τ.

With the formally exact expression (4.68) for the generating function and equation (4.54), any observable can, in principle, be calculated. In reality, since this is a problem with exponentially growing complexity, this can mostly never—neither analytically nor numerically—be done without further simplifications. When used properly, however, these expressions serve as the foundation for a successful numerical treatment, as we show in the following section.

4.2. The ISPI Method

In recent years, a number of methods were developed that allow to study the dynamics of a nonequilibrium system like the SLQD in contact with metallic wires and go beyond linear response or perturbation theory of first and second order. For example, if the highly correlated stationary state of a quantum dot with strong on-site interactions is of interest, the nonequilibrium Bethe ansatz [122–124] may be a suitable method to consider. In case that the real-time dynamics is to be studied, worthwhile candidates for theoretical frameworks can surely be found among such diverse schemes as the various renormalisation group methods (functional, numerical, ...) [125–127, 129, 130, 133–135, 139–141, 196, 197], the flow-equations approach [128], or real-time quantum Monte-Carlo methods (rtQMC) [146, 148, 149]. At last, we can also include the ISPI approach, as developed by Weiss et al. [120, 198, 199], into this list, which is by no means complete. Each of these schemes has its specific characteristics, advantages, and shortcomings—sometimes they complement each other, sometimes their scope of applications overlap.[11]

Before we introduce it in detail below, we briefly outline the specifics of ISPI as compared to some of the above mentioned methods. ISPI is based on a formally exact, discrete path integral representation of the Keldysh partition function. All approximations we introduce below to further simplify this expression come with parameters that allow to adjust the degree of approximation (quasi-)continuously, starting from zero (the exact expression). Therefore ISPI is numerically exact, as opposed to, e.g., renormalisation group approaches. It allows to describe the real-time dynamics of the SLQD in contact with the metallic leads, which distinguishes it

---

[11]For more details regarding other existing theoretical approaches, see chapter 1.

4. Iterative Summation of Path Integrals

from stationary approaches like the Bethe ansatz, for example. It is built upon the central assumption that the interaction with the continuous bath restricts the life-time of correlations to some finite value as long as not both the temperature T and bias voltage V_bias are zero or very small. In cases, when this assumption is valid (essentially, outside the deep Kondo regime), ISPI permits to study long propagation times (unlike rtQMC, due to the fermionic sign problem and the dynamic phase problem), while still taking into account all relevant correlations. It is specifically designed to describe a system, for which all appearing energy scales, particularly the interaction parameters U and J, are of the same order of magnitude as the tunnel coupling Γ, which we choose as reference unit for the energy scale. This is in striking contrast to the perturbative approach of chapter 3, where Γ had to be much smaller than all dot energies.

Of course, in addition to the already mentioned case of $T = V_\text{bias} = 0$, there are applications, systems as well as regimes, to which ISPI cannot be applied. The fact that ISPI involves the complete summation of a path integral—the reason why it does not suffer from the fermionic sign problem—leads to a high numerical complexity that grows exponentially with the model size and restricts applications to rather small interacting systems. As we discuss at the end of section 4.2.4, the SLQD with Coulomb interaction and magnetic impurity is already at the limit (in size) for systems that are treatable with today's computational capacities and reasonable computing times. Also, very short propagation times (below $\sim 1\hbar/\Gamma$) lie in principle well within the scope of the ISPI method—their calculation does, in fact, not even require an iterative evaluation of the generating function. The present formulation of the ISPI scheme, which employs a particular method of cancelling divergent terms in the inverse Green's function (shown in section 4.2.3), however, is not suitable to handle short propaga-

4. Iterative Summation of Path Integrals

tion times.[12] Finally, since the path integral is basically an expansion in orders of the Coulomb- and electron-impurity interactions, ISPI only converges well in the regime with small to intermediate interaction strength ("few" Γ), when not "too much" expansion orders are relevant.

In the present section, we review the iterative summation of the (fermionic) path integral and adopt it to the model of the single-level quantum dot with the fixed, spin-1/2 magnetic impurity (2.1b) that interacts with electrons on the dot via exchange interaction (2.1c). While the iterative scheme for the calculation of $\mathcal{Z}[\eta]$ is explained, mainly according to reference [120], particular attention is paid to the specific modifications necessary to account for the magnetic impurity and its interaction with the electrons on the quantum dot, since this is a new extension of existing methods. The section is closed with the introduction of the extrapolation procedures to numerically eliminate systematic errors that are due to the necessary time discretization and the assumed finite coherence time.

4.2.1. Finite Correlation Length Approach

In its full complexity, the exact generating function (4.68) is unserviceable to us, as its exact evaluation for long propagation times[13] would require calculating determinants of two rather large matrices (and one inversion) for every instance of an exponentially growing number of paths $\{\tau\}$ and $\{\zeta\}$. For example, the numerical evaluation of expression (4.68) with a ratio of time interval and discretization step of around $\Delta_t/\delta_t \approx 100$, a value not unusual for the calculations presented below, would involve around 4^{100} determinants of 100-dimensional matrices. Obviously, this is far beyond the reach of any existing computer system. Even with one of the state-of-the-

---

[12]Since ISPI is not even needed in these cases, this is not a very important issue at this point.

[13]As "long" we characterise time differences Δ_t with $\Delta_t \Gamma / \hbar \gtrsim 10$

4. Iterative Summation of Path Integrals

art QMC methods [146, 148, 149], this kind of calculation would not yield results with acceptable statistical errors in reasonable time. Hence, in order to utilise $\mathcal{Z}[\eta]$ for our purposes, we cannot proceed without an additional, suitable numerical treatment.

In this context, it is worthwhile to study the contribution of long time correlations to the system dynamics. Off-diagonal elements of the inverse Green's matrix $(G^{\text{eff}})^{-1}_{t,t'}$ are closely related to time non-local correlations, i.e., their absolute values for some time difference $\Delta t := t - t'$ indicate how strongly the system "remembers," at a given time, its state over Δt in the past. This relation is actually oblique, as it is the Green's function itself and not its inverse that gives the two-particle correlations. In section 4.2.3 and appendix G, however, we show that G and G^{-1} behave very similar as far as elements at large Δt are concerned. To account for *every* matrix element of $(G^{\text{eff}})^{-1}$ in the evaluation of $\mathcal{Z}[\eta]$ for arbitrary long propagation times is therefore tantamount to include all possible time correlations indiscriminately. Yet, in many physical systems, particularly if they can be described by quantum statistics, one can expect that correlations prevail only over a certain (short) period, which we will call memory- or coherence time τ_c and which is determined by the dynamical conditions. If this is the case, correlations rapidly die off and can be neglected as soon as Δt exceeds τ_c. In case of model (2.1), the off-diagonal elements of the inverse Green's function are given by the lead gamma function $\gamma(p, t - t')$ in expression (4.62). It follows directly, that all components of this matrix show in fact a fast, even exponential decay with increasing distance from the diagonal, as can be seen in figure 4.5. In addition to that, a non-zero bias voltage results in an oscillatory behaviour of the real and imaginary parts of each component. Since this decay is induced by the coupling to the leads alone, it can only be given in terms of lead parameters. As it turns out, these parameters are the (inverse) temperature T and (to a minor degree) the bias voltage V_{bias}.

4. Iterative Summation of Path Integrals

While the temperature is directly responsible for the exponential decay, a growing V_{bias} is believed to lead to an increasingly effective cancellation of long time correlations due to oscillations.

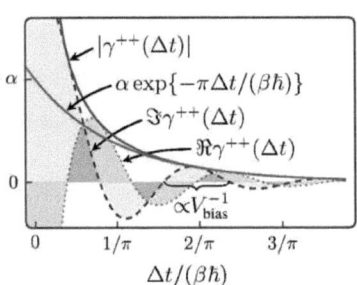

Figure 4.5.: Time dependence of the lead gamma matrix with $\alpha = \Gamma/(2\beta\hbar^2)$, $eV_{\text{bias}}\beta = 4\pi$, and $\Delta t := t - t'$. Shown are the absolute value and both the real and imaginary part. For $\pi|\Delta t|/(\beta\hbar) \gtrsim 2$, the absolute value of $\gamma^{++}(\Delta t)$ decays exponentially, with a decay time proportional to the inverse temperature. Both the real and imaginary part are oscillatory functions, whose periods are essentially given by the bias voltage. Since $\gamma(\Delta t)$ is closely related to correlations in the system, T and V_{bias} determine the lead induced memory time.

Thus, matrix elements of $(G^{\text{eff}})^{-1}$ further away from the diagonal than τ_c do not (considerably) contribute to $\mathcal{Z}[\eta]$. In other words, it is the key observation that all correlations beyond τ_c are not relevant for the system's behaviour, while the effective value of τ_c is determined via the numerical memory convergence procedure that is described in section 4.2.4. This motivates the following approximation: with the effective coherence time $\tau_c =: (K-1)\delta_t$ given, all matrix elements $(G^{\text{eff}})^{-1}_{ij}$ with $|i-j| \geq K$ are set to zero. Depending on the propagation time $t_f - t_i$, this can drastically reduce the numerical costs for the evaluation of $\mathcal{Z}[\eta]$. It should be noted, however, that the appropriate value of τ_c is not identical to the decay time of the gamma matrix and generally not easy to obtain. This is due to the non-trivial way, in which correlations enter the generating function, and complicated further by the fact that both T and V_{bias} influence the decay time of γ. Rather, in the ISPI scheme, K is used as a numerical parameter in an extrapolation procedure that yields results that are independent of

4. Iterative Summation of Path Integrals

the memory time[14] but can be obtained with minimised numerical efforts. It is introduced in section 4.2.4. Both steps together, the dropping of the matrix elements beyond τ_c and the extrapolation to infinite memory time, constitute the *finite correlation range approach* (FCA).

At first, to exploit the FCA, it is helpful to rearrange the elements of the Green's matrix. The Green's matrix has a 2×2 Keldysh block structure, where each block has a real-time substructure. With appropriate (identical) permutations of lines and columns, these structures can be interchanged. The determinant of the matrix is left unaltered by this operation. This is illustrated by the following example

$$\begin{pmatrix} a_{11}^{++} & a_{12}^{++} & a_{13}^{++} & a_{11}^{+-} & a_{12}^{+-} & a_{13}^{+-} \\ a_{21}^{++} & a_{22}^{++} & a_{23}^{++} & a_{21}^{+-} & a_{22}^{+-} & a_{23}^{+-} \\ a_{31}^{++} & a_{32}^{++} & a_{33}^{++} & a_{31}^{+-} & a_{32}^{+-} & a_{33}^{+-} \\ a_{11}^{-+} & a_{12}^{-+} & a_{13}^{-+} & a_{11}^{--} & a_{12}^{--} & a_{13}^{--} \\ a_{21}^{-+} & a_{22}^{-+} & a_{23}^{-+} & a_{21}^{--} & a_{22}^{--} & a_{23}^{--} \\ a_{31}^{-+} & a_{32}^{-+} & a_{33}^{-+} & a_{31}^{--} & a_{32}^{--} & a_{33}^{--} \end{pmatrix} \mapsto \begin{pmatrix} a_{11}^{++} & a_{11}^{+-} & a_{12}^{++} & a_{12}^{+-} & a_{13}^{++} & a_{13}^{+-} \\ a_{11}^{-+} & a_{11}^{--} & a_{12}^{-+} & a_{12}^{--} & a_{13}^{-+} & a_{13}^{--} \\ a_{21}^{++} & a_{21}^{+-} & a_{22}^{++} & a_{22}^{+-} & a_{23}^{++} & a_{23}^{+-} \\ a_{21}^{-+} & a_{21}^{--} & a_{22}^{-+} & a_{22}^{--} & a_{23}^{-+} & a_{23}^{--} \\ a_{31}^{++} & a_{31}^{+-} & a_{32}^{++} & a_{32}^{+-} & a_{33}^{++} & a_{33}^{+-} \\ a_{31}^{-+} & a_{31}^{--} & a_{32}^{-+} & a_{32}^{--} & a_{33}^{-+} & a_{33}^{--} \end{pmatrix},$$
(4.70)

where the lines indicate the Keldysh superstructure before and the real-time structure after the transformation, respectively. When they are ascendingly numbered from 1 to 6, the permutation that was applied to both the lines and columns can be identified with (2 3 5 4) in cycle notation. The hatched elements are the ones that would be set to zero for $K = 2$, i.e., $\tau_c = \delta_t$. While they are scattered all over the matrix in the conventional Keldysh notation, after the rearrangement, they form solid blocks a distance (counted in blocks) away from the main diagonal that corresponds to the difference of their element's real-time step indices. In the general case, the effective

---

[14]Or, equivalently, incorporate all relevant correlations.

4. Iterative Summation of Path Integrals

inverse Green's matrix for a given K can be written as

$$(G_\sigma^{\text{eff}})^{-1}_{kl} \approx \begin{pmatrix} \boxed{A}^\sigma_{1,1} & \boxed{A}^\sigma_{1,2} & & \\ \boxed{A}^\sigma_{2,1} & \boxed{A}^\sigma_{2,2} & \ddots & \\ & \ddots & \ddots & \boxed{A}^\sigma_{N_c-1,N_c} \\ & & \boxed{A}^\sigma_{N_c,N_c-1} & \boxed{A}^\sigma_{N_c,N_c} \end{pmatrix}, \qquad (4.71)$$

where $N_c = N/K$ and the blocks $\boxed{A}^\sigma_{k,l}$ are K dimensional matrices, whose entries are filled with elements of $(G_\sigma^{\text{eff}})^{-1}$ from rows and columns in the range of $\{(\imath-1)K+1,\ldots,\imath K\}$ with $\imath = k,l$. Due to the approximation, an $\boxed{A}$-block in one of the secondary diagonals is strictly upper (lower) triangular, if it is below (above) the main diagonal. The Ξ_σ matrices in expression (4.68) inherit the same block structure as the approximate inverse Green's matrix exhibits in equation (4.71), with the exception that the to $\boxed{A}^\sigma_{k,l}$ corresponding blocks $\boxed{B}^\sigma_{k,l}$ of a Ξ_σ matrix are, in general, not quadratic as their dimensions depend on the number of flip-flop processes within the respective coherence time interval.

This is exemplified in figure 4.6. The presented (discrete) impurity path of length $8\delta_t$ consists of $2N = 18$ vertices and features 12 flip-flop processes. Also shown is the approximate matrix $\Xi_\downarrow$ for $\tau_c = 2\delta_t$, corresponding to $K = 3$, that is obtained after the interchange of superstructures and setting $G^{\alpha\alpha'}_{\downarrow,qr} = 0$ for $|q - r| \geq K$ (hatched matrix elements). The path with $N = 9$ can then be divided into $N_c = 3$ disjunct path segments with K vertices on each branch. In analogy to the blocks $\boxed{A}^\downarrow_{i,i}$, the diagonal matrix blocks $\boxed{B}^\downarrow_{i,i}$ then contain all matrix elements $G^{\alpha\alpha'}_{\downarrow,qr}$ with $(i-1)K < q, r \leq iK$ (dashed outlined boxes).

If the number of flip-flops in some of the segments is odd, or, more specifically, the number of creator- and annihilator fields associated to one seg-

4. Iterative Summation of Path Integrals

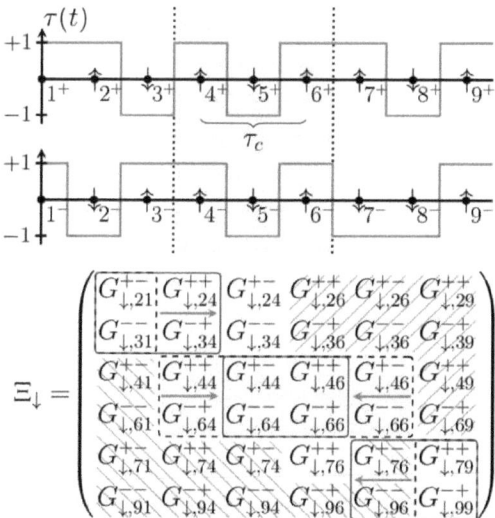

Figure 4.6.: Exemplary impurity path (above) with 12 flip-flops and the corresponding approxi- mate $\Xi_\downarrow$-matrix (below) for $\tau_c = 2\delta_t$ ($K = 3$). The *discrete* path (arrows on vertices) has a length of $8\delta_t$ ($N = 9$) and can be divided into $N_c = 3$ disjunct segments of length K (separated by dotted lines). Depending on the distribution of flips, the number of creator- and annihilator fields assigned to the different segments may differ. In that case, the diagonal blocks $\boxed{B}_{i,i}^{\downarrow}$ are not quadratic (dashed outlined boxes) and their determinants do not exist. As a consequence, the $\boxed{B}^{\downarrow}$ have to be modified for the iterative calculation of $\det \Xi_\downarrow$. This is done by an appropriate re-assignment (blue arrows) of fields that belong to flips closest the segment borders, so that all diagonal blocks become quadratic (solid boxes). Hatched elements are set to zero.

ment i do not coincide, the matrix block $\boxed{B}_{i,i}^{\downarrow}$ is not quadratic. In the example of figure 4.6, this is the case for all three segments. The fields $\mathfrak{d}_{\downarrow,1}^-$, $\mathfrak{d}_{\downarrow,2}^+$, and $\mathfrak{d}_{\downarrow,3}^-$ belong to the first, fields $\mathfrak{d}_{\downarrow,4}^+$, $\bar{\mathfrak{d}}_{\downarrow,4}^+$, $\bar{\mathfrak{d}}_{\downarrow,4}^-$, $\bar{\mathfrak{d}}_{\downarrow,6}^+$, $\mathfrak{d}_{\downarrow,6}^+$, $\mathfrak{d}_{\downarrow,6}^-$ to the second segment, and the remaining fields $\mathfrak{d}_{\downarrow,7}^+$, $\bar{\mathfrak{d}}_{\downarrow,9}^+$, $\mathfrak{d}_{\downarrow,9}^-$ belong to segment three. Hence, this construction scheme yields 2×1-matrices $\boxed{B}_{1,1}^{\downarrow}$ and $\boxed{B}_{2,2}^{\downarrow}$, and

4. Iterative Summation of Path Integrals

a 2×4-matrix $\boxed{B}_{3,3}^{\downarrow}$. As we show in the next section, however, the iterative calculation of the generating function requires the *diagonal* blocks to be quadratic, so that their determinant can be calculated. This can be achieved by re-assigning the earliest and latest creator fields $\bar{\mathfrak{d}}_{\downarrow,4}^{+}$ and $\bar{\mathfrak{d}}_{\downarrow,6}^{-}$ from the second segment to the first and third, respectively. In doing so, all diagonal blocks become quadratic (blue arrows). This kind of reshaping of the $\boxed{B}_{i,i}^{\sigma}$-blocks is possible for all paths. Therefore, from now on, we assume that all diagonal blocks of the approximate Ξ_σ-matrices are quadratic.

4.2.2. The Iterative Scheme

With the finite correlation length approach, the matrices $(G_\sigma^{\text{eff}})^{-1}$ and Ξ_σ in equation (4.68) can be written in block form, where all blocks besides those on the main or neighbouring secondary diagonals vanish [see equation (4.71)]. The reshaping of the $\boxed{B}_{i,i}^{\sigma}$-blocks, where necessary, ensures that we are dealing with matrices, whose diagonal blocks are quadratic and invertible. For the further considerations, it is convenient to introduce a recursive notation to describe such a matrix $\mathbf{X}$ with block dimension D:

$$\mathbf{X} = \boxed{\boxed{\mathbf{X}_D}} \text{ with } \boxed{\boxed{\mathbf{X}_{D-i+1}}} := \begin{pmatrix} \boxed{X}_{i,i} & \boxed{X}_{i+1,i} \\ \boxed{X}_{i,i+1} & \boxed{\boxed{\mathbf{X}_{D-i}}} \end{pmatrix} \text{ and } \boxed{\boxed{\mathbf{X}_1}} = \boxed{X}_{D,D}.$$
(4.72)

Whole block matrices are doubly outlined, where the subscript gives their dimension in blocks. The determinant of $\mathbf{X}$ can be calculated iteratively (or recursively) within $D-1$ iterations, while each iteration of the procedure consists of three steps:

(4.I1) Perform one step of Gaussian elimination to get rid of the block in the second row, first column.

4. Iterative Summation of Path Integrals

(4.I2) In the resulting element of second row, second column, neglect products like $[X]_{k-1,k}[X]_{k,k+1}$, which connect path segments with indices differing by more than one.

(4.I3) Expand the determinant after the first column, thus reducing the problem by one in block dimension.

The first and third step actually suffice to yield the exact value of the determinant for a block matrix like (4.71). In this sense, they are mere algebraic operations. Step (4.I2), however, is different insofar as it introduces an additional approximation, necessary to fully exploit the computational benefits of the FCA. With its help, an iteration step $k \to k + 1$ can be performed solely based on the knowledge of (i) the determinant after step k and (ii) the spins orientations in segments k and $k + 1$. This is because the inclusion of terms that connect segments with indices differing by more than one, would require information about impurity spins in segments $< k$. Relative to those spins in segment $k + 1$, these are farther than τ_c in the past. Thus, by the additional approximation (4.I2), terms that are relatively small due to a coupling of spins farther apart than τ_c are disregarded. It is consistent with the FCA and we show in appendix G that it can indeed be expected to yield only small deviations from the "exact" value, while considerably reducing the numerical costs of the method.

We can now derive a general iterative expression for the determinant of $\mathbf{X}$. Operation (4.I1), i.e., the first step of Gaussian elimination, is performed by multiplying the first row of $\mathbf{X}$ with $-[X]_{2,1}[X]_{1,1}^{-1}$ and adding it to the second row. Since in this first iteration no higher-order products are added

4. Iterative Summation of Path Integrals

to $\boxed{X}_{2,2}$, we can proceed with step (4.13) to arrive at

$$\det \mathbf{X} = \det \boxed{X}_{1,1} \det \begin{pmatrix} \boxed{X}'_{2,2} & \boxed{X}_{2,3} \\ \boxed{X}_{3,2} & \boxed{\mathbf{X}_{D-2}} \end{pmatrix} \tag{4.73}$$

where $\boxed{X}'_{2,2} = \boxed{X}_{2,2} - \boxed{X}_{2,1}\boxed{X}_{1,1}^{-1}\boxed{X}_{1,2}$.

With this, we are almost done as the next iteration steps are largely identical and the iterative expression for the determinant is formally given by $\det \mathbf{X} = \det \boxed{X}_{1,1} \prod_{i=2}^{D} \det \boxed{X}'_{i,i}$. We just have to fix the general expression for the iterated, modified diagonal blocks $\boxed{X}'_{i,i}$, while considering iteration step (4.12). This can be done with mathematical induction (see appendix G) and yields an expression for the diagonal blocks that is completely analogous to (4.73) for all $i \geq 2$. We obtain the final iterative identity

$$\det \mathbf{X} = \det \boxed{X}_{1,1} \prod_{i=2}^{D} \det\{\overbrace{\boxed{X}_{i,i} - \boxed{X}_{i,i-1}\boxed{X}_{i-1,i-1}^{-1}\boxed{X}_{i-1,i}}^{\boxed{X}'_{i,i}}\}. \tag{4.74}$$

At this point, we have to address the question of how, besides the above mentioned reshaping, to properly construct an approximate version of the Ξ_σ-matrices for finite coherence time. In the exact version of the generating function (4.68), they have to be filled with elements of the *full* Green's matrix $G_\sigma^{\text{eff}}[\{\tau, \varsigma\}]$, which depends on the whole spin path $\{\tau, \varsigma\}$ between t_i and t_f. Formally, it can be calculated by inverting the full inverse Green's function, a numerically very costly procedure for a typical number of $N \sim 100$ time steps that would render the previous approximation steps useless. Rather, in conformity with the FCA, we have to find approximate values for the elements of blocks $\boxed{B}^\sigma_{i,i(\pm 1)}$ that solely depend on spins of segments

4. Iterative Summation of Path Integrals

neighbouring $i(\pm 1)$. Again, the basic idea for finding such an approximation is based on the assumption that blocks on the secondary diagonals contribute less to any matrix expression (inverse, determinant, etc.) than those on the main diagonal. The easiest way to make use of this assumption is to expand the inverse of $(G_\sigma^{\text{eff}})^{-1}$ in orders of the non-diagonal blocks. This straightforward procedure, which is shown in appendix G, results in

$$\boxed{G}_{k,k} \approx \boxed{A}_{k,k}^{-1} \quad \text{and} \quad \boxed{G}_{k,l} \approx -\boxed{A}_{k,k}^{-1}\boxed{A}_{k,l}\boxed{A}_{k,l}^{-1} \qquad (4.75)$$

for all $1 \leq k,l \leq N_c$ and with $|k-l| = 1$ (the electron spin index was omitted). The blocks $\boxed{G}_{k,l}$ are defined in analogy to the $\boxed{A}$-blocks of the inverse Green's function. From the approximate expressions on the r.h.s., the $\boxed{B}$-blocks are filled as described above. Now, we possess (almost) all ingredients needed for calculating the generating function iteratively from the knowledge of the inverse Green's function.

4.2.3. Cancelling of Divergent Terms in the Generating Function

Before we can write down the iterative generating function, however, we have to deal with the $(t-t')^{-1}$ singularity of the lead's contribution $(G_{\text{env}}^0)^{-1}$ to the inverse Green's function. There is an elegant way to get rid of the divergent elements in all the matrices that appear in the generating function, while preserving the exponential decay with growing distance to the diagonal. It was introduced by Weiss et al. [120] and makes use of the fact that, in contrast to its inverse, the Green's function itself is finite at $t = t'$ (or, more precisely, its limits for $t' \to t$ are). From a mathematical point of view, the idea is to analytically cancel the divergent, spin-path independent

4. Iterative Summation of Path Integrals

factor in the fraction

$$\frac{\mathcal{Z}[\eta] - \mathcal{Z}[0]}{\eta \mathcal{Z}[0]} \approx i\langle O \rangle(t_{\text{EV}}), \quad (4.76)$$

which is the numerical expression used to evaluate the expectation value (4.54). In other words, if the divergent part $\det G_{\text{div}}^{-1}$ of $\mathcal{Z}$ does (i) neither depend on the paths of impurity- and HS-spin nor (ii) the small parameter η, while (iii) its inverse is finite for all values of t, t', we can divide the generating function by the singular factor without affecting the expectation value $\langle O \rangle(t_{\text{EV}})$. Ultimately, we would require that the FCA is also applicable to the resulting generating function $\mathcal{Z}'[\eta]$.

Fortunately, all these conditions are fulfilled for our model system. The $(t - t')^{-1}$ singularity is contained in the term $(G_{\text{env}}^0)^{-1}$, which originates from the dot-lead-coupling and, hence, is independent both of $\tau(t)$ and $\zeta(t)$. Note that since τ and ζ are directly associated to the system's interactions, all terms that do not depend on them describe the non-interacting (free) part of the system. Obviously, $(G_{\text{env}}^0)^{-1}$ does not depend on η as well and even in case of $\hat{O} = \hat{I}$, the corresponding source term (4.63) is finite. As it turns out, to meet condition (iii) it is necessary and, as we will see, convenient to identify G_{div}^{-1} with the sum of *all* non-interacting, source independent contributions, viz., the free inverse electronic Green's function:

$$(G_{\text{div}})_{t,t'}^{-1} \equiv \sum_\sigma (G_{0,\sigma}^{\text{el}})_{t,t'}^{-1} = \sum_\sigma (G_{\text{dot},\sigma}^{\text{el},0})_{t,t'}^{-1} + (G_{\text{env},\sigma}^0)_{t,t'}^{-1} \quad (4.77)$$

with

$$(G_{\text{env},\sigma}^0)_{t,t'}^{-1} = \sum_p \gamma(p, t - t'), \quad (4.78)$$

$$(G_{\text{dot},\sigma}^{\text{el},0})_{t,t'}^{-1} = \delta(t - t') \begin{pmatrix} i\partial_t - \omega_\sigma^U & 0 \\ 0 & -i\partial_t + \omega_\sigma^U \end{pmatrix}, \quad (4.79)$$

4. Iterative Summation of Path Integrals

with $\omega_\sigma^U := (\epsilon_\sigma + U/2)/\hbar$. Since the generating function factorises into two independent parts for each electron spin orientation σ, it is sufficient to deal with the corresponding $(G_{0,\sigma}^{\text{el}})_{t,t'}^{-1}$ only. Before its inverse is calculated, we Fourier-transform it into frequency space, where it is a finite matrix function. This yields

$$(G_{0,\sigma}^{\text{el}})^{-1}(\omega) = \begin{pmatrix} \omega - \omega_\sigma^U + i\Gamma/\hbar[F(\omega) - 1] & -i\Gamma/\hbar F(\omega) \\ i\Gamma/\hbar[2 - F(\omega)] & -\omega + \omega_\sigma^U + i\Gamma/\hbar[F(\omega) - 1] \end{pmatrix}. \quad (4.80)$$

We defined $F(\omega) := f_{\text{L}}(\omega) + f_{\text{R}}(\omega)$ and $(G_{0,\sigma}^{\text{el}})_{\omega,\omega'}^{-1} =: 2\pi\delta(\omega - \omega') \times$ $\times (G_{0,\sigma}^{\text{el}})^{-1}(\omega)$. The matrix is obviously finite for all ω, ω' as long as the coupling Γ does not vanish and can be inverted algebraically. The result is then transformed back into time space by means of complex contour integration. The whole inversion process is shown in appendix G, where the rather lengthy expressions for the free Green's function in time space (rather, for its dimensionless version $\tilde{G}$) are presented in equation (G.18).

Its essential features are illustrated with the help of its $++$ component in figure 4.7. All components of $(\tilde{G}_{0,\sigma}^{\text{el}})_{t,t'}$ only depend on the difference $t - t'$ and decay exponentially with growing $|t - t'|$, where non-zero values of the dot frequency ω_σ^U or the bias voltage eV_{bias} lead to an oscillating behaviour (blue line and shading). Compared to the γ-matrix, the temperature dependence, though still present, is only weakly pronounced and cannot be read from the figure. The real part of the Green's function in figure 4.7(a) is smooth and finite on the whole real-time axis and shows no singularities. The imaginary part shares these characteristics with the exception of the point $\Delta t = 0$, where it shows a discontinuous jump and is not well-defined. However, since both the (left and right) limits $t' \to t \pm 0^+$ of $\Im(\tilde{G}_{0,\sigma}^{\text{el}})_{t,t'}$ exist, this singularity turns out to be unprob-

4. Iterative Summation of Path Integrals

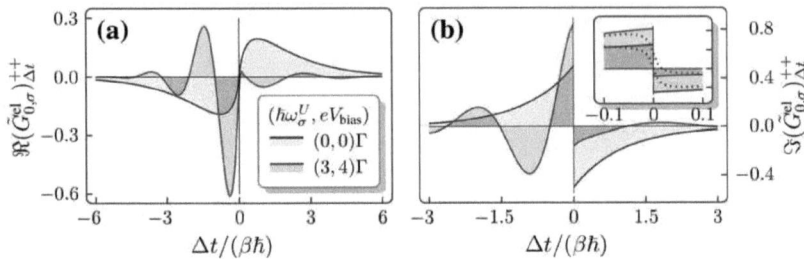

Figure 4.7.: The ++ component of the dimensionless free Green's function $\tilde{G}^{\text{el}}_{0,\sigma}$, as given by equation (G.18), versus time difference $\Delta t = t - t'$ for two combinations of ω^U_σ and eV_{bias} and with $\beta\Gamma = 1/5$. The absolute value decays exponentially with growing $|\Delta t|$ and both real part **(a)** and imaginary part **(b)** oscillate, if ω^U_σ or eV_{bias} are non-zero. With exception of the point $\Delta t = 0$, the free Green's function is smooth and finite on the whole real-time axis and shows no divergent behaviour. The imaginary part jumps discontinuously at zero time difference and is not defined for $\Delta t = 0$. For our further calculations and in contrast to the divergence of the γ-matrix, however, this singularity turns out to be harmless and there a several ways to account for it properly (see text). One possible option is to introduce a high frequency cutoff term $\exp\{|\omega|/\omega_c\}$ with $\omega_c > 0$ to the *inverse* Green's function in equation (4.2.3) before inserting it and transforming it back into time space. The inset [vertical scaling identical to main figure (b)] shows the result of this procedure (dotted lines) with the exact function (solid) for $\omega_c = 100\Gamma$.

lematic for the following considerations. Before this point is analysed in detail, we cancel out the divergence factor in the generating function (4.68) by multiplying by $-i \det G^{\text{el}}_{0,\sigma}$, or equivalently, replacing $i(G^{\text{eff}}_\sigma)^{-1}$ with

$$D_\sigma[\eta] := G^{\text{el}}_{0,\sigma}(G^{\text{eff}}_\sigma)^{-1} = \mathbb{1} + G^{\text{el}}_{0,\sigma}(\sigma\Sigma^0_\sigma + \eta\Sigma^\eta_\sigma),$$

(4.81)

where $(\Sigma^0_\sigma)_{kl} = \begin{pmatrix} -J\tau_k/\hbar - i\zeta_k\lambda(\delta_t) & 0 \\ 0 & J\tau_k/\hbar - i\zeta_k\lambda^*(\delta_t) \end{pmatrix} \delta_{kl}\delta_t.$

4. Iterative Summation of Path Integrals

The value of Σ_σ^η depends on the observable $\hat{O}$. In case of $\hat{O} = \hat{I}$, we can identify it with $(G_{\text{env}}^I)^{-1}$ as given by equation (4.63) and with $(G^O)^{-1}$, otherwise.

The next step is to properly discretise the new matrix D_σ or, rather, the matrices $G_{0,\sigma}^{\text{el}}$ and Σ_σ^η, while accounting for the discontinuity of the free Green's matrix. Due to the way how the continuous inverse Green's functions were defined and to the fact that integrations over time carry a time unit, some attention has to be paid to the proper definition of this discretization. In appendix G, we lay out that a discrete matrix X_{kl} with corresponding time step δ_t can be consistently obtained from a continuous function $X(t,t')$ according to

$$X_{kl} := \int_{t_k-\delta_t/2}^{t_k+\delta_t/2} dt \int_{t_l-\delta_t/2}^{t_l+\delta_t/2} dt'\, X(t,t') \quad \text{with} \quad t_k = t_1 + (k-1)\delta_t. \quad (4.82)$$

Apparently, this requires that the continuous version of a discrete, dimensionless matrix in real-time space carries the dimension of time^{-2} (note that the Dirac delta distribution has dimension time^{-1}). For the free Green's matrix, this yields

$$(G_{0,\sigma}^{\text{el}})_{kl} = \frac{1}{\delta_t^2} \int_{t_k-\delta_t/2}^{t_k+\delta_t/2} dt \int_{t_l-\delta_t/2}^{t_l+\delta_t/2} dt'\, (\tilde{G}_{0,\sigma}^{\text{el}})_{t,t'} \approx (\tilde{G}_{0,\sigma}^{\text{el}})_{t_k,t_l}, \quad (4.83)$$

where the approximate relation is valid as long as the integrand is continuous and the time step δ_t small enough [see equation (G.18) for a definition of $(\tilde{G}_{0,\sigma}^{\text{el}})_{t,t'}$].

Around the discontinuity of the free Green's function in case of $k = l$, however, the approximation does not hold, since in this point the r.h.s. is not defined. The integral on the l.h.s., though, still yields a proper discrete value for $(G_{0,\sigma}^{\text{el}})_{k,k}$, since the integrand is finite and both left and right limits

4. Iterative Summation of Path Integrals

exist. Thus, an easy and consistent way to get rid of this singularity is to define the diagonal elements of the free Green's matrix as

$$(G^{\text{el}}_{0,\sigma})_{kk} := [(\tilde{G}^{\text{el}}_{0,\sigma})_{\Delta t \to 0^-} + (\tilde{G}^{\text{el}}_{0,\sigma})_{\Delta t \to 0^+}]/2, \quad (4.84)$$

i.e., as the average of both limits. It can be seen from the inset in figure 4.7(b) that this value coincides well with the one obtained by introducing a high frequency cuttoff to the Green's function in energy space—the method used by Weiss et al. [120] to deal with the singularity. To illustrate this, we numerically evaluated the real-time Green's function after multiplying expression (4.2.3) with a factor $\exp\{|\omega|/\omega_c\}$, where $\omega_c = 100\Gamma$. For $\omega^U_\sigma = V_{\text{bias}} = 0$, both methods result in a zero diagonal value. For $\omega^U_\sigma = 3\Gamma$ and $eV_{\text{bias}} = 4\Gamma$, equation (4.84) yields $(G^{\text{el}}_{0,\sigma})_{k,k} \approx 0.3384i$ and with the cutoff method we obtain $0.3245i$. The relative difference is about 4% and gets smaller for larger values of ω_c. In the end, both ways are consistent and compatible, but the use of equation (4.84) has two advantages: it does not alter non-diagonal values of the Green's function whatsoever and avoids the introduction of an additional parameter ω_c.

For all the observables we consider in this work, the discrete version of Σ^η_σ can be obtained from equation (4.82) in a straightforward manner. In case of the charge current, for example, we plug in expression (4.63) and get

$$\left(\Sigma^\eta_\sigma\right)_{kl} = \frac{e\Gamma\delta_t}{2\beta\hbar^2} \frac{\sin[eV_{\text{bias}}(k-l)\delta_t/(2\hbar)]}{\sinh[\pi(k-l)\delta_t/(\hbar\beta)]} \begin{pmatrix} \delta_{km} - \delta_{lm} & -\delta_{km} \\ \delta_{lm} & 0 \end{pmatrix}, \quad (4.85)$$

where we set $t_{\text{EV}} = t_m = t_1 + (m-1)\delta_t$.

With the expressions (4.81), (4.83), and (4.85), we have all ingredients needed to write down a divergence-free generating function that nevertheless exponentially decays with increasing $|k-l|$. The FCA can be applied

4. Iterative Summation of Path Integrals

to the discrete matrix $D_\sigma[\eta]_{kl}$, which is then re-ordered according to equation (4.70). As shown above, the result is a block matrix, whose blocks $\boxed{D}^\sigma_{i,i(\pm 1)}$ are used to iteratively calculate $\mathcal{Z}'[\eta]$ (see below). Yet, before we can proceed, it has to be considered how the construction of the Ξ-matrices is affected by the transition from $(G^{\text{eff}}_\sigma)^{-1}$ to D_σ. The equations (4.75), which were introduced in the previous section, involve blocks of the inverse Green's function and are difficult to handle numerically due to the divergence at $\Delta t = 0$. Therefore, our goal is to use only the well-behaved matrices D_σ and $G^{\text{el}}_{0,\sigma}$ to construct the Ξ-matrices. From equation (4.81), we can immediately derive that $G^{\text{eff}}_\sigma = D_\sigma^{-1} G^{\text{el}}_{0,\sigma}$. We evaluate this by using equation (4.75) to obtain the approximate inverse of D_σ and multiply the result with the free Green's matrix in block form. For iteration step $k-1 \to k$, these steps yield (see appendix G for details)

$$\begin{aligned}\boxed{G}_{x,x} &= \boxed{D}^{-1}_{x,x}\left\{\boxed{G_0}_{x,x} - \boxed{D}_{x,\bar{x}}\boxed{D}^{-1}_{\bar{x},\bar{x}}\boxed{G_0}_{\bar{x},x}\right\} \\ \boxed{G}_{x,\bar{x}} &= \boxed{D}^{-1}_{x,x}\left\{\boxed{G_0}_{x,\bar{x}} - \boxed{D}_{x,\bar{x}}\boxed{D}^{-1}_{\bar{x},\bar{x}}\boxed{G_0}_{\bar{x},\bar{x}}\right\},\end{aligned} \quad (4.86)$$

where we defined the index $x \in \{k-1, k\}$ with $\bar{x} = k - \delta_{k,x}$. With this, the $\boxed{B}$-blocks can be constructed without the inverse Green's function and we can now write $\mathcal{Z}'[\eta]$ iteratively as

$$\mathcal{Z}'_{(\delta_t, K)}[\eta] = \sum_{\{\tau,\zeta\}_{N_c}} \mathcal{Z}'_{N_c} \quad \text{with} \quad \mathcal{Z}'_{i>1} = \sum_{\{\tau,\zeta\}_{i-1}} \Lambda_{i,i-1} \mathcal{Z}'_{i-1}, \quad (4.87a)$$

$$\Lambda_{i,i-1} = F_i \prod_\sigma \prod_{X=B,D} \det\{\boxed{X}^\sigma_{i,i} - \boxed{X}^\sigma_{i,i-1}\boxed{X}^{\sigma,-1}_{i-1,i-1}\boxed{X}^\sigma_{i-1,i}\}, \quad (4.87b)$$

and

$$\mathcal{Z}'_1 := F_1 \prod_\sigma \prod_{X=B,D} \det \boxed{X}^\sigma_{1,1}. \quad (4.87c)$$

Here, we defined $\{\tau,\zeta\}_i = \{\tau\}_i \cup \{\zeta\}_i$ with $\{s\}_i := (s^\mp_{iK}, \ldots, s^\mp_{(i-1)K+1})$

157

4. Iterative Summation of Path Integrals

and $s = \tau, \zeta$ as (sub-)tuple of those impurity- and HS-spins that lie in the i-th path segment of length K (see figure 4.8). Related to the number and position of flip-flop processes is the prefactor

$$F_i = 2^{-2K}(-1)^{m_i^-}\left(\frac{J\delta_t}{2\hbar}\right)^{m_i} e^{i\Phi_{\mathrm{imp}}^{(i)}}, \tag{4.88}$$

where

$$\Phi_{\mathrm{imp}}^{(i)} = -\frac{\Delta_{\mathrm{imp}}\delta_t}{2\hbar}\sum_{j=(i-1)K+1}^{iK}(\tau_j^+ - \tau_j^-),$$

and m_i is the number of flip-flops in segment i, of which m_i^- lie on the backwards branch. The factor 2^{-2K} can as well be omitted, since it cancels out when evaluating an observable with the help of equation (4.76). The subscripts of the generating function in equation (4.87a) are there to point out that $\mathcal{Z}'_{(\delta_t,K)}[\eta]$ is not exact, but afflicted with systematic errors that depend on two approximations and their associated parameters: (i) the step width δ_t of the time discretization and (ii) the FCA with coherence time $\tau_c = (K-1)\delta_t$. In the limit of infinitely small δ_t and infinitely large K, however, the iterative expression becomes exact again:

$$\mathcal{Z}'[\eta] = \lim_{K\to\infty}\lim_{\delta_t\to 0}\mathcal{Z}'_{(\delta_t,K)}[\eta]. \tag{4.89}$$

4.2.4. Extrapolation Procedures

At first glance, it may seem that compared to the already exact version (4.68), there is not much gained with the iterative expression (4.89). A choice of both very small time steps δ_t and large values of K—the numerical equivalent of the exact limit—results in large dimensions of the matrix blocks and, more importantly, an exponentially large number of spin paths to sum over.

4. Iterative Summation of Path Integrals

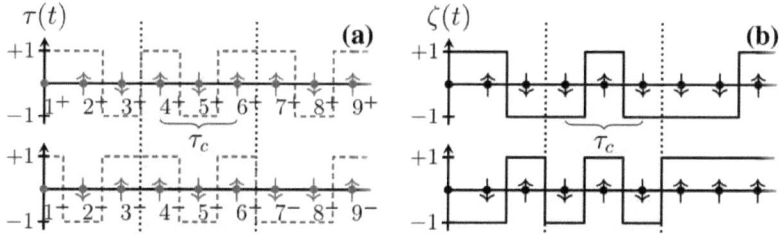

Figure 4.8.: Examplary discrete paths (tuples) for the impurity- [gray in subfigure **(a)**] and HS field [black in **(b)**]. For $\tau_c = 2\delta_t$ ($K = 3$) both paths break down into segments $\{s = \tau, \zeta\}_i$ of length $K = 3$ in real-time, which contain $2K = 6$ spins (K per contour branch). For example, path segment $\{\tau\}_1 = (\tau_3^-, \tau_3^+, \tau_2^-, \tau_2^+, \tau_1^-, \tau_1^+) = (\uparrow, \downarrow, \downarrow, \uparrow, \uparrow, \uparrow)$ is the tuple that contains the 6 "real-time earliest" impurity spins. Accordingly, we find that $\{\zeta\}_3 = (\zeta_9^-, \zeta_9^+, \zeta_8^-, \zeta_8^+, \zeta_7^-, \zeta_7^+) = (\uparrow, \uparrow, \uparrow, \downarrow, \uparrow, \downarrow)$. For short, we defined segments of combined impurity- and HS spins as $\{\tau, \zeta\}_i = \{\tau\}_i \cup \{\zeta\}_i$. For the second segment, we can write $\{\tau, \zeta\}_2 = (\uparrow, \uparrow, \downarrow, \downarrow, \uparrow, \uparrow, \downarrow, \downarrow, \uparrow, \uparrow, \downarrow, \downarrow)$

Nevertheless, in cases when the FCA is valid for our model system, there exists a finite, yet unknown, physical (as opposed to its numerical analogon) memory for time correlations and it can be expected that convergence to the exact limit is fast, once the numerical parameter τ_c is of the same order of magnitude as this "true"(physical) memory. Depending on the other system parameters, this can be the case for values of K that are accessible to a numerical treatment. In this section, we present criteria, how to decide if such a situation is at hand and, if so, use them to obtain extrapolated values of $\mathcal{Z}'[\eta]$ that are independent of δ_t and K. These values, although still afflicted with some numerical errors, can then be regarded as (quasi-)exact. Accordingly, the extrapolation procedures can be regarded as a numerical equivalent of evaluating the analytical limits in equation (4.89).

Once the extrapolation procedures are established, the iterative scheme as given by equation (4.87), can unfold its full potential. It is apparent that

4. Iterative Summation of Path Integrals

all the iteration steps given in equation (4.87) have the same numerical complexity. Since the system propagates by one coherence time τ_c with each iteration, this means that the numerical costs of the ISPI procedure scale linearly with evolution time $t_{\text{EV}} - t_i$. As we explained in the introduction to this section, this is one of the major benefits of the ISPI method.

Prior to discussing the extrapolation, however, we have to assess the question of reasonable upper and lower bounds for the choice of the numerical parameters δ_t and K, respectively. For the time step, we already found two limiting conditions: in the course of the Hubbard-Stratonovich transformation in section 4.1.4, the condition $\hbar/\delta_t \geq U/\pi$ was derived, while the term $[J\delta_t/(2\hbar)]^{m_i}$ in prefactor F_i [equation (4.88)] suggests that $\hbar/\delta_t$ should at least be greater than $J/2$. As a generalisation of these results, we require that the upper limit for the time step should be related to the maximum of the system's energies:

$$\hbar/\delta_t \gtrsim \max(\Phi_D, \Delta, \Delta_{\text{imp}}, U, J/2, eV_{\text{bias}}, k_{\text{B}}T). \tag{4.90}$$

As far as the energies are considered that do not characterise an interaction (all except U and J), this is just a rough estimation. In case of the bias voltage, for example, it turns out that even at $eV_{\text{bias}} = 4\Gamma$ reliable results can be obtained for $\hbar/\delta_t = 2\Gamma$.

An estimation of the lower bound for the coherence time parameter τ_c can be gained from the study of the exponential decay of the elements of D_σ. These are essentially determined by the free Green's function and the γ-matrix, as shown in figures 4.5 and 4.7. However, since the time correlations can just be limited by the influence of the reservoirs, only three parameters come into consideration, anyway: the temperature T, the bias voltage V_{bias}, and the coupling Γ. Obviously, without a tunnel coupling correlations in the dot would never die out. On the other hand, the whole

4. Iterative Summation of Path Integrals

transport scheme rests on a finite coupling, which we chose as reference value for all appearing energies. This manifests itself in the fact, that the free Green's function essentially decays as $\exp\{-\Gamma/\hbar\,|\Delta t|\}$ [see equation (G.18)]. As a consequence, this leaves only the lead parameters β and V_{bias} as candidates, which are relevant for the estimate of τ_c:

$$\tau_c \simeq \hbar \min(|eV_{\text{bias}}|^{-1}, \beta). \tag{4.91}$$

We want to stress, that connected to this relation is the requirement that *not both* the temperature and the bias voltage are zero or very small. Although long time correlations cancel out in that case as well, they do so only algebraically (for $T = V_{\text{bias}} = 0$). As a consequence, the numerical errors of our method will increase gradually for lower and lower values of both T and V_{bias}. For all following considerations, we implicitly assume that δ_t and τ_c are chosen in agreement with (4.90) and (4.91).

Extrapolating to the Limit of Vanishing Time Step

Due to the Trotter break-up of the time contour (the time discretization), the path integral and, in consequence, all expressions that are based on it are, by construction, exact only up to terms that scale with δ_t^2 (see section 4.1). This precise knowledge about the expected behaviour of deviations from the exact value, in other words, the systematic errors, can be used to formulate a strong theoretical criterion for being close to convergence.

(4.C1) For one fixed τ_c and several values $\delta_t^{(1)} < \delta_t^{(2)} < \ldots < \delta_t^{(n)}$ of the discretization step, closeness to the exact, "true" value $\langle O \rangle(\tau_c)$ of some observable is measured by how well the results $\langle O \rangle(\tau_c, \delta_t^{(1)})$, $\langle O \rangle(\tau_c, \delta_t^{(2)}), \ldots, \langle O \rangle(\tau_c, \delta_t^{(n)})$ are described by linear function of δ_t^2.

4. Iterative Summation of Path Integrals

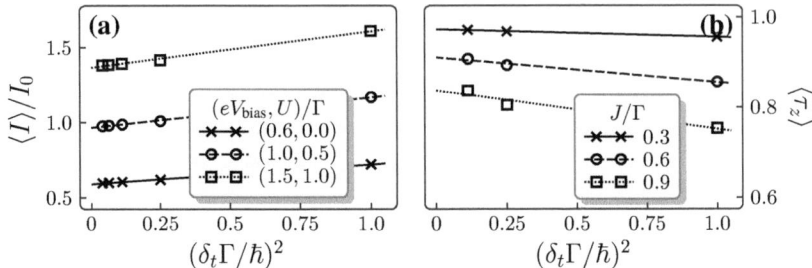

Figure 4.9.: Graphical illustration of the extrapolation procedure to (numerically) obtain expectation values in the limit $\delta_t \to 0$ and with finite τ_c. For all plots, we chose $\tau_c = \hbar/\Gamma$, $t_{\rm EV}\Gamma/\hbar = 4$, $\beta\Gamma = 1$, and $\Phi_D = \Delta = \Delta_{\rm imp} = 0$. When plotted against $(\delta_t\Gamma/\hbar)^2$, the values (crosses, circles, and squares) are fitted to linear functions (solid lines). A small relative standard deviation from the linear behaviour indicates that unsystematic errors are small compared to the systematical trotter error. In that case, it is safe to extrapolate to $\delta_t = 0$ and treat this value as exact limit. **(a)** Charge current in units of $I_0 = e\Gamma/\hbar$ for $J = 0$. **(b)** Expectation value $\langle \tau_z \rangle$ of the impurity spin orientation for $eV_{\rm bias} = 0.5\Gamma$ and $U = 0.5\Gamma$. The impurity was initially in the spin-up state [$\tau_i = 1$ in (4.48)].

In figure 4.9 is shown how this criterion is used for extrapolation to the limit $\delta_t \to 0$. For a fixed value of $\tau_c = \hbar/\Gamma$ and several differing values of δ_t (corresponding to consecutive numbers $K = 2, 3, \ldots$), we calculated the charge current $\langle I \rangle$ [figure part (a)] and the impurity spin orientation $\langle \tau_z \rangle$ [part (b)] at evaluation time $t_{\rm EV} = 4\hbar/\Gamma$ for temperature $\beta\Gamma = 1$ and with $\Phi_D = \Delta = \Delta_{\rm imp} = 0$. For each configuration of *physical* parameters (as opposed to the numerical parameters δ_t and τ_c), we plot the calculated expectation values (plot marks) against $(\delta_t\Gamma/\hbar)^2$ and fit them to a linear function (solid lines). The smaller the relative error of this fit is, the better criterion (4.C1) is fulfilled by the numerical results. In all plots of 4.9(a), the errors (based on the sample standard deviation) are below 1%, while in sub-figure (b) they grow from around 1% for $J = 0.3\Gamma$ to about 10% for $J = 0.9\Gamma$. From this, we can infer that the *unsystematic* numerical error, which results in randomly distributed deviations from the "true" calculation result and allows to estimate the numerical stability of the ISPI implementation for a certain set of parameters, is relatively small. It is mainly given by the sample standard deviation from the theoretically known behaviour of the *systematic* errors due to the time discretisation (a linear function of δ_t^2) and has to be propagated for the extrapolation to infinite memory times. Another unsystematic error arises from the numerical derivative of the generating function (as given by equation 4.76) and manifests in small imaginary parts of the observable. Its relative size ranges typically between 10^{-5} to 10^{-3} and is in most cases at least one order of magnitude smaller than the deviations of the numerical values from the linear fit. It can therefore mostly be neglected. This is also true for the results in figure 4.9, where the relative deviations from the linear behaviour are small and we can safely take the value at $\delta_t = 0$ of the fitted function as the numerical (quasi-)exact limit for infinitely small time step (though, with a somewhat lower reliability for $\langle \tau_z \rangle$ at $J = 0.9\Gamma$).

Extrapolating to the Limit of Infinite Memory Time

After the extrapolation to $\delta_t \to 0$, the numerically obtained expectation values $\langle O \rangle(\tau_c)$ are independent of the step width of the Trotter break-up δ_t. What remains is to get rid, by some additional extrapolation, of the τ_c dependence as well. Compared to the removal of the Trotter error, however, this turns out to stand on less solid theoretical grounds, as a strict mathematical dependence of the systematic error on the value of the finite memory time is unknown. Therefore, the criteria we present here rather have the sta-

4. Iterative Summation of Path Integrals

tus of empirical observations that were established by a combination of (i) comparison with analytical results and (ii) reasons of epistemic simplicity (see below), while their interpretation involves physical intuition and experience using the numerical approach. With this said, we formulate (and subsequently discuss) the following convergence criteria.

(4.C2) Several differing values $\langle O \rangle(\tau_c)$ indicate convergent behaviour, if

   a) they show a *linear* dependence on $1/\tau_c$ with relatively small mean deviations.

   b) their dependence on $1/\tau_c$ can be described by a reasonably smooth and "flat" function that exhibits a local extremum (principle of least dependence, see also [52, 198–200])

Apparently, criterion (4.C2.b) is a weaker indicator of convergence than is the first, (4.C2.a), which should be generally preferred. Several qualifiers such as "relatively small" or "reasonably flat" are not quantifiable strictly and for general cases but have to be interpreted in the context of a particular situation, in the light of one's experience and physical intuition, and, if possible, adjusted by comparison to known results, which were obtained by other means. Nevertheless, though being "empirical", both criteria should not be regarded as mere hand-waving guidelines to obtain a limit $\tau_c \to \infty$, at will. Not only were they already successfully applied by [52, 200] (for the QUAPI method, which is closely related to ISPI), Weiss et al. [120, 198, 199], who were able to reproduce important well-known results, e.g. the Landauer-Büttiker current [201, 202], for a SLQD in contact with metallic leads. Aside from that, there are also theoretical arguments for the utility and validity of (4.C2).

Since we are interested in the limit $\tau_c \to \infty$, it is an obvious choice to study how the numerically obtained expectation values $\langle O \rangle$ depend on

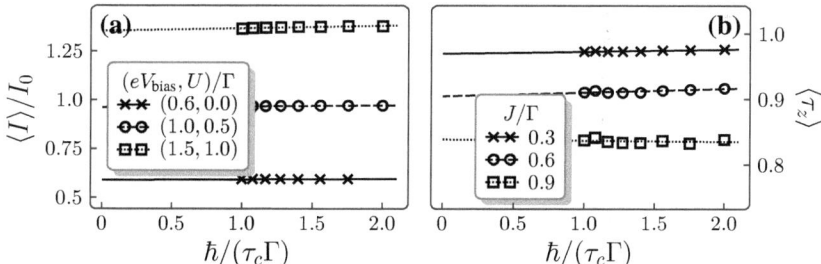

Figure 4.10.: Graphical illustration of the extrapolation to the limit $\tau_c \to \infty$. If the values $\langle O\rangle(\tau_c^{-1})$ (crosses, circles, and squares) are well-described by a linear function, it is assumed that the numerical parameter τ_c is close to or larger than the "true" memory time of the system and that the linear behaviour continues until $\tau_c^{-1} = 0$. **(a)** Extrapolation of the charge current. For all three combinations of eV_{bias} and U the (relative) standard deviations to the linear fit are below 1%. **(b)** Impurity orientation. The (relative) standard deviations for the linear fits are well below 1%.

the *inverse* powers of τ_c. Without any a priori knowledge, a linear relation in τ_c^{-1}, as demanded by (4.C2.a), is the simplest, non-constant functional dependence to assume. Close to convergence, we expect that $\langle O\rangle(\tau_c^{-1})$ does not change strongly and reaches a stationary value at $\tau_c^{-1} = 0$. Close enough to this limit, if it is not already constant, any smooth function of the inverse coherence time is approximately linear. Hence, if values $\langle O\rangle(\tau_c^{-1})$ for decreasing τ_c^{-1} can be fitted well to a linear function, we consider this to be a reliable convergence indicator and extrapolate to $\tau_c^{-1} = 0$. It is the underlying assumption, that in this case the numerical parameter τ_c is of the same order of magnitude or bigger than the "true", yet unknown, physical coherence time and, therefore, the linear behaviour will persist until infinity. This is illustrated in figure 4.10 both for the current and the impurity spin in sub-figure (a) and (b), respectively, where the physical parameters correspond to those in figure 4.9.

4. Iterative Summation of Path Integrals

Criterion (4.C2.b) applies to a more complex situation. It comes into play, when the $\langle O \rangle (\tau_c^{-1})$, while still showing a relatively weak dependence on the inverse memory time (the function is "flat"), are—by virtue of featuring a local extremum—not described well by a linear function. Such a behaviour is the result of a trade-off between accuracy and computational costs that has to be found, when choosing the parameters δ_t and K. The minimisation of the Trotter error requires as small time steps δ_t as possible. At the same time, an as large as possible coherence time τ_c is needed to optimally account for the bath correlations. Naturally, these optimisations with respect to accuracy are limited by the quickly increasing numerical costs. For a certain range of K values, it may be possible to decrease both δ_t and τ_c^{-1} simultaneously and, in doing so, obtain more accurate results. Yet, at some point, a further increase of the memory time eventually requires to choose larger time steps in order to keep running times of the simulations in check. This will, in turn, lead to a larger Trotter error and, as a consequence, a decrease of the accuracy. It can also cause the function $\langle O \rangle (\tau_c^{-1})$ to feature a local extremum. Suppose, for example, that for some choice of parameters, the values $\langle O \rangle (\tau_c^{-1})$ tend to fall both for decreasing δ_t and increasing τ_c. Hence, as long as the computational costs allow choosing larger K without the need to increase δ_t, the value of $\langle O \rangle (\tau_c^{-1})$ would decrease. As soon as δ_t has to be enlarged to limit the running time of the simulation, however, the decrease of $\langle O \rangle$ due to larger K can be over-compensated by an increase due to larger δ_t, thus, resulting in a local minimum. As in the linear case, a "weak" dependence on τ_c^{-1} motivates the assumption that hopefully "not much" will happen when approaching $\tau_c^{-1} = 0$ and that convergence is close.

In cases like this, the *principle of least dependence* (PLD) proves to be useful. The whole idea of the extrapolation to infinite τ_c (which does not work well in the outlined situation) is to eliminate any dependence of $\langle O \rangle$ on the memory time. Inspired by this fact, the principle of least dependence suggests to choose that value as approximate limit $\tau_c \to \infty$ of $\langle O \rangle(\tau_c^{-1})$, which *depends least* on τ_c. Apparently, this value is identical to the local extremum. By definition, this choice has the benefit of being most stable, i.e., independent of the choice of fit function. In the spirit of the previous paragraph, it also is a palpable candidate for the best compromise between accuracy and running time. To actually implement this principle into our method, though, we will not fit any function to the calculated expectation values, but just pick the value closest to the extremum. This is illustrated in figure 4.11(a). For $\Phi_D = -\Gamma/2$, $\Delta = \Gamma/2$, $J = -\Gamma/2$, $\beta\Gamma = 5$, $\Delta_{\text{imp}} = U = 0$, and four different propagation times, the calculated values show a local minimum for coherence times between $\tau_c \Gamma/\hbar = 0.5$ and 1.[15] With increasing evaluation times t_{EV}, the variations of $\langle \tau_z \rangle(\tau_c^{-1})$ grow as well. In the figure, a different scaling of the vertical axis for each t_{EV} emphasises these variations by showing them in relation to the respective mean value. All plots share the same lower limit of the scale: $\tau_z^{\min} = 0.025$, but have different upper limits $\tau_z^{\max}$. For example, the red curve for $t_{\text{EV}} = 44\hbar/\Gamma$ is plotted for $\tau_z^{\max} = 0.32$. The PLD is applied by just picking the smallest from each set of values for a given t_{EV} as limit $\tau_c \to \infty$. In sub-figure (b), the results of this "extrapolation" are shown for propagation times between $t_{\text{EV}}\Gamma/\hbar = 0$ and 80 with $\langle \tau_z \rangle(0) \equiv \tau_i = 1$. They are well described by an exponentially decaying function (solid line, standard deviation ~ 0.005)— in agreement with the physically expected behaviour (see next chapter).

To recapitulate, the conditions (4.C2), on which both these extrapolation procedures rely, are empirical, as is the underlying assumption that τ_c exceeds the physical memory time, once these criteria are met. Furthermore, they require to be interpreted with experience and with respect to the par-

---

[15] The fitted functions (solid lines) are just a guide to the eye.

ticular physical situation. For example, with increasing values of the interaction parameters—especially at low temperatures—statistical errors grow rapidly and, at some point, render the extrapolation schemes useless or at least doubtful. Yet, whether and when this point is reached, is not always easy to decide, and it would surely be advantageous to have a better physical understanding of how the expectation values depend on τ_c to improve the extrapolation. Nevertheless, the two procedures of obtaining values in the limit of infinite memory time yield, under the right conditions, very accurate results that agree well with outcomes that have been obtained either analytically or by numerical methods with overlapping scopes of application, where available (see the next chapter and [52, 120, 198–200]).

Restricting the Number of Flip-Flops per Memory Time

Let us close this chapter with some remarks about the computational costs of the ISPI method and a way to reduce them. Since the number of spin paths to sum over at each iteration step grows exponentially with τ_c, the running time of the numerical simulation mainly depends on the parameter K. For vanishing electron-impurity interaction ($J = 0$), i.e., in the absence of flip-flop processes, the number of paths grows essentially like 2^K, as only the constant path $(\tau_i, \ldots, \tau_i)$ contributes to a sum in (4.87). In this case, simulations with values of $2 \leq K \leq 6$ with reasonable running times of up to a few days can be carried out with our sequential version of the code.[16] As shown in figure 4.9(a), this allows to produce up to 5 data points for an extrapolation to $\delta_t \to 0$—more than enough in almost all cases.

When the electron-impurity interaction is non-zero, the computational costs are considerably higher. The number of paths to sum up then grows like 4^K, leading to running times of over a month for values as small as

---

[16] A parallelized version is possible but requires a strongly refined focus on numerical procedures.

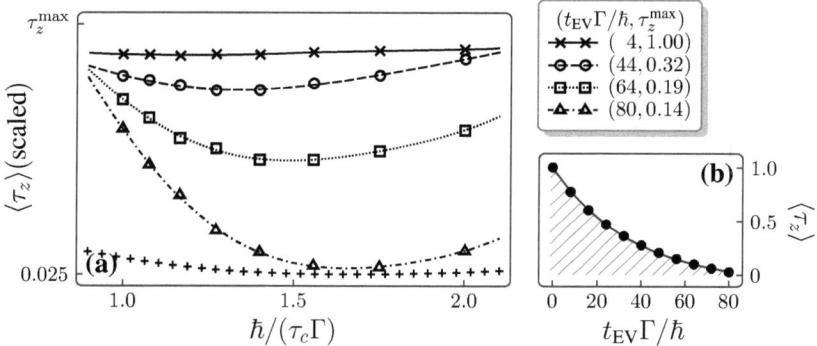

Figure 4.11.: (a) Impurity spin orientation against τ_c^{-1} for four different values of evaluation time t_{EV}. The vertical axis is scaled differently for each time to emphasise the relative size of variations. For all plots, the lower bound of the scale is set to $\tau_z^{\min} = 0.025$, while the upper bound $\tau_z^{\max}$ is given in the plot legend. As model parameters, we chose $\Phi_D = -\Gamma/2$, $\Delta = \Gamma/2$, $J = -\Gamma/2$, $\beta\Gamma = 5$, and $\Delta_{\text{imp}} = U = 0$. The lines, polynomial fits, are guides to the eye. The line of '+' marks corresponds to the dashdotted plot for $t_{EV}\Gamma/\hbar = 80$, when scaled like the solid line for $t_{EV}\Gamma/\hbar = 4$. Since all plots show a local minimum, we pick from each set of values the smallest one as best estimate of the limit $\tau_c \to \infty$ of $\langle \tau_z \rangle$. This implements the principle of least dependence (PLD, see text). (b) $\langle \tau_z \rangle$ as function of t_{EV}, where all values were obtained using the PLD. If it was initially in the spin-up state, the impurity orientation is expected to decay exponentially (see chapter 5). The calculated values confirm this expectation, as they deviate only slightly (standard deviation 0.005) from an exponential fit curve (solid line).

$K = 4$ (comparable in complexity to $K = 8$ for a SLQD without impurity). However, to decide whether and how well the simulation results show a linear dependence on δ_t^2, necessary for the extrapolation to $\delta_t = 0$, we need at least three data points, corresponding to $K = 2, 3, 4$. In cases, when the "true" coherence time in the system is too large to obtain a reliable result for $\delta_t = \tau_c$ ($K = 2$), a value for $\delta_t = \tau_c/4$ ($K = 5$) would even be needed. On contemporary computers, this can not be calculated in reasonable time,

4. Iterative Summation of Path Integrals

at least not without a strongly parallelized code.

Fortunately, there is a way to drastically reduce the running times for $J \neq 0$, while affecting the final results—in most cases—only within the bounds of the numerical errors. It is based on the observation that the path-dependent prefactor F_i in equation (4.88) contains a small number $[J\delta_t/(2\hbar)]$ to the power of flip-flops $0 \leq m_i \leq 2(K-1)$ in path segment i. Hence, an individual path segment contributes the less to a sum in equation (4.87) the higher its number of flip-flops. At the same time, the number of impurity path segments $\{\tau\}_i$ with m_i flip-flops—given by $4C_{m_i}^{2(K-1)}$ with $C_k^n = n!/[k!(n-k)!]$—grows for m_i between 0 and $K-1$ and *decreases again* for $k \leq m_i \leq 2(K-1)$. Together with the observation that, aside from the prefactor, each path contribution is roughly of the same order of magnitude, this motivates the following assumption.

(4.T1) Depending on the observable $\hat{O}$ as well as the model parameters, particularly J, δ_t, and K, there exists a $m_i^{\max} < 2(K-1)$, so that all contributions from (impurity spin) paths with $m_i > m_i^{\max}$ can be neglected in the iterative scheme.

For increasing number of flip-flops per path segment, the rapidly decreasing weight of individual paths due to prefactor $[J\delta_t/(2\hbar)]^{m_i}$, may be (over-)compensated, at first, by their increasing numbers for $0 \leq m_i \leq K-1$. Considering the smallness of $[J\delta_t/(2\hbar)]$, however, and by taking into account that the number of paths decreases for even larger m_i, it is reasonable to assume that the prefactor cannot be compensated for all classes of paths. At the latest, when the number of flip-flops is very close to the maximum $2(K-1)$, this is easy to see: Both the path classes with $m_i = 0$ and $m_i = 2(K-1)$ contain the same number of elements (four), while each contribution from a path in the second class comes with a prefactor

4. Iterative Summation of Path Integrals

$[J\delta_t/(2\hbar)]^{2(K-1)}$. For typical values of $K = 4$, $\delta_t\Gamma/\hbar = 1/2$, and $J = \Gamma$ it amounts to $\sim 2.5 \times 10^{-4}$. Hence, it is very likely that the value of any observable would not change within the accuracy bounds of the numerical method if we neglected the path class with $m_i = 2(K-1)$. In many situations, in fact, more than just the path class with the highest flip-number can be neglected. However, since we cannot present a theoretical estimation of $m_i^{\max}$ that is valid for general cases, we just included it into our code as additional parameter. Then we performed a numerical estimation by means of a spot sample of the parameter space. It turns out, that for many parameter configurations and $K = 4$, good results can already be obtained with $m_i^{\max} = 2$. This can reduce the running times from over a month to a few days. Figure 4.12 shows an example of such a "sample point." For $K = 4$ and increasing numbers of flip-flops, both $\langle I \rangle$ and $\langle \tau_z \rangle$ converge rather quickly to the value that is produced by the full path sum.

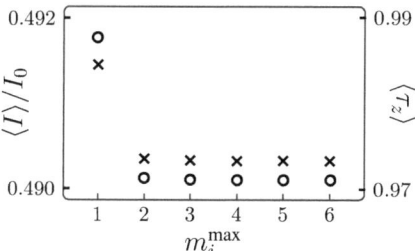

Figure 4.12.: Current (crosses, left scale) and impurity interaction (circles, right scale) for $t_{\text{EV}}\Gamma/\hbar = 4$, $\beta\Gamma = 1$, $\Phi_D = \Delta = \Delta_{\text{imp}} = 0$, $\delta_t\Gamma/\hbar = 1/3$, $K = 4$, and increasing values of $m_i^{\max} = 1, \ldots, 6$. It can be seen, that both for $\langle I \rangle$ and $\langle \tau_z \rangle$, the values for $m_i^{\max} = 2$ already coincide well with the result for the full path summation with $m_i^{\max} = 2(K-1) = 6$. The running time, which for the latter amounts to more than one month, is reduced to a few days if paths with three and more flip-flops are neglected.

4.3. Summary

In this chapter, we have derived a fermionic path integral representation of the Keldysh partition function for the SLQD with spin-1/2 magnetic impurity, from which a generating function was constructed. Based on this generating function, we reviewed how the numerically exact ISPI method can be used to calculate expectation values of observables $\hat{O}$ by neglecting long-time correlations as it is known that they are exponentially suppressed. In doing so, we had to adopt and extend the method that was originally developed for a simple SLQD to the full model (2.1) with impurity. This required to find a proper short time propagator, which allowed to construct the generating function $\mathcal{Z}[\eta]$ [see equation (4.68)] for arbitrary observables, based on the mixed basis of fermionic coherent states for the electrons and the usual, discrete spin states $|\tau =\uparrow, \downarrow\rangle$ for the impurity. The final expression captured the effect of the non-trivial two-body interactions (Coulomb- and flip-flip scattering) in the path sum over discrete spin-1/2 fields ($\{\tau, \zeta\}$), while all other degrees of freedom could be traced out analytically.

Compared to the SLQD without impurity, the electron-impurity interaction not only manifests itself in an additional spin field to sum over numerically, which already effectively doubles the *speed of* (exponential) *growth* of computing times with increasing interactions or system size. Furthermore, for each impurity path with flip-flops the determinants of additional matrices Ξ_τ, which contain elements of the full electronic Green's function, have to be evaluated. Since its exact values are not known, we had to find an approximation of the Green's function in order to calculate $\mathcal{Z}[\eta]$. This approximation had to be consistent with the finite correlation length approach, while still yielding the (formally) exact result for $\tau_c \to \infty$. What is more, as diagonal blocks of Ξ_τ are not necessarily quadratic for all possible path segments (spin tuples of length $2K$), we developed and implemented rules

4. Iterative Summation of Path Integrals

to consistently reshape these blocks, so that the iterative scheme of path summation can be applied.

Finally, we discussed extrapolation procedures to eliminate the systematic errors of the iterative scheme due to finite time step δ_t and memory time $\tau_c = (K-1)\delta_t$. In case of $J \neq 0$, the high numerical costs for even the most moderate extrapolation to $\delta_t = 0$ made it necessary to devise an additional approximation to reduce running times of the simulation for $K = 4$ and higher. It is based on the observation that classes of impurity paths generally contribute the less to the path sum the more flip-flop events they feature. To implement this procedure, we had to devise an algorithm that builds, for a given maximum number $m_i^{\max}$ of flip-flops per coherence time, tables of compatible path segments, ensuring that for neighbouring segments no subpath of length K has more than $m_i^{\max}$ flip-flops.

As can be seen from the results in the previous section, the iterative summation of path integrals is a powerful numerical scheme to calculate the real-time evolution of a SLQD—with or without impurity—in the deep nonequilibrium quantum regime. The non-perturbative, exact method is applicable to a parameter regime, where all (interaction) energies are of the same order of magnitude and accounts for all relevant correlations in the system. Since the computational complexity scales linearly with the propagation time, very long times of $80\hbar/\Gamma$ and more can be reached [see figure 4.11(b)]. In the next chapter, we will use the ISPI method to investigate the dynamics of model system (2.1), focusing on the stationary current and relaxation behaviour of the impurity spin.

5

Spin-Relaxation by a Charge Current: Exact Results

THE ISPI SCHEME was originally developed for the Coulomb-interacting single-level quantum dot (SLQD). By reproducing well-known and, where available, experimentally confirmed results that were described either analytically or with the help of established theoretical methods, Weiss et al. [120] were able to verify the general validity of the method. Based on that, further steps into previously unexplored parameter ranges and towards potentially new physics have been made, as well. We are not going to review these (published) results here, but focus our attention mainly on those novel transport features that are introduced to the model along with the interacting impurity.

In this context, it is crucial to notice, that new dynamics and transport behaviour can only arise due to the transverse- or flip-flop interaction $\hat{H}_{\text{int}}^{\perp}$, as given by equation (4.5). Without the possibility for flip-flops, the orientation of the impurity spin and, by extension, its quantum state could not change and not participate in the dynamics. What is more, the remaining longitudinal part of the interaction $\hat{H}_{\text{int}}^{\parallel}$ would then only account for a constant energy shift of the electron levels—an effect that could equivalently be achieved by the application of a magnetic field. With the Zeeman energy

5. Spin-Relaxation by a Charge Current: Exact Results

Δ, this situation is already accounted for in the model without impurity. Hence, to see non-trivial dynamics that follows from the presence of the impurity spin, we have to consider the effects of the electron-impurity flip-flops. The simplest phenomenon that we can study in this regard is the time dependence of the impurity orientation $\langle \tau_z \rangle$.

5.1. A Rate Equation for the Impurity Spin

Figure 5.1 shows the time evolution of $\langle \tau_z \rangle$ for different values of the interaction parameter J, where the impurity was set to be in the spin-up state initially.[1] For the remaining model parameters, we choose $\Phi_D = \Delta = \Delta_{\text{imp}} = 0$, $U = \Gamma/2$, $\beta\Gamma = 1$, and $eV_{\text{bias}} = 0.6\Gamma$. The impurity spin orientation, which we also call polarisation, shows an exponentially decaying behaviour for intermediate to long propagation times. Not surprisingly, the polarisation decays the faster, the stronger the impurity interacts with the electron spins. The decay can be explained by a rate equation ansatz, similar to the one shortly reviewed in section 2.3. In doing so, we assume that the impurity orientation in contact with a "bath" of the tunnelling electrons is governed by a Markovian dynamics and can be described solely by the time dependent probabilities $P_\tau(t)$ of finding the impurity in state $|\tau\rangle$ at time t, which are given by the nonequilibrium Bloch equations

$$\dot{P}_\tau(t) = W_\tau P_{-\tau}(t) - W_{-\tau} P_\tau(t) \quad \text{with} \quad P_\uparrow(t) + P_\downarrow(t) = 1. \qquad (5.1)$$

The time independent W_τ are effective rates for transitions (effective spin flips, see below) from state $|-\tau\rangle$ to state $|\tau\rangle$. We can then identify $\langle \tau_z \rangle(t) = P_\uparrow(t) - P_\downarrow(t)$ and solve equation (5.1) in a straightforward manner (for de-

---
[1]This is equivalent to setting $\tau_i = 1$ in equation (4.48), which will be used as initial condition throughout the rest of the chapter.

5. Spin-Relaxation by a Charge Current: Exact Results

tails, see appendix H) to obtain

$$\langle\tau_z\rangle(t) = \tau_i \, e^{-2\overline{W}(t-t_i)} + \frac{w}{2\overline{W}}\left(1 - e^{-2\overline{W}(t-t_i)}\right), \tag{5.2}$$

where we defined the average rate $\overline{W} := (W_+ + W_-)/2$ and the difference in rates $w := W_+ - W_-$. As usual, τ_i denotes the initial impurity orientation. For the chosen parameters the model system is isotropic, i.e., symmetric with respect to (relative) spin orientations. Furthermore, the anti-ferromagnetic interaction favours anti-parallel orientation of electron- and impurity spin. Over long propagation times, the coupling to the unpolarised leads then also destroys any polarisation of the impurity. It follows immediately, that in those cases the rates for up- and down flips are equal as well, and, for $\tau_i = 1$, we get

$$\langle\tau_z\rangle(t) = e^{-2W(t-t_i)} \tag{5.3}$$

from equation (5.2), where we set $W_+ = W_- =: W$. Apparently, this simple theoretical prediction agrees well with the numerical results from the figure.

This can be ascribed to the choice of parameter range we study in this section. While in this regime the interaction energies of the system are all of the same order of magnitude and allow for a considerable influence of correlated effects on the transport behaviour, the rather high temperature and bias voltage nevertheless reduce the relevance of coherent dynamics to an overall secondary role. This is beneficial for the purpose of this chapter, which is to establish the validity of the ISPI method for the description of the exact impurity dynamics of and the stationary current through our version of a magnetic Anderson model. On the one hand, we show below that the exact numerical results will be in sufficient accordance with a simple

5. Spin-Relaxation by a Charge Current: Exact Results

perturbative theory to obtain a first intuitive explanation of the impurity dynamics and "transfer" its plausibility to the ISPI results. On the other hand, coherence-induced deviations from the simple theory are large enough to clearly illustrate the need for a non-perturbative theoretical description.

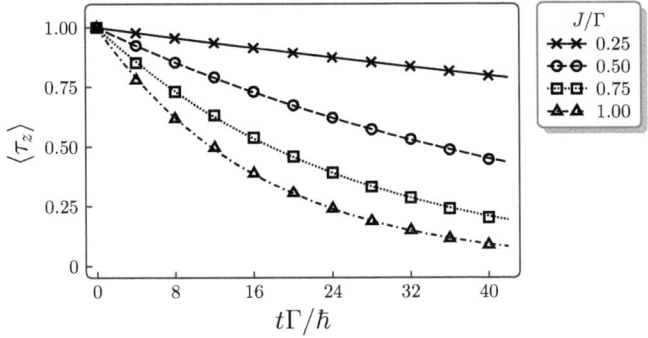

Figure 5.1.: The expectation value $\langle \tau_z \rangle$ of the impurity orientation as a function of time for four different strengths J of an anti-ferromagnetic electron-impurity interaction. The system was initially "prepared" in a spin-up state $[\tau_z(0) \equiv \tau_i = 1]$ and we set $\Phi_D = \Delta = \Delta_{\text{imp}} = 0$, $U = \Gamma/2$, $\beta\Gamma = 1$, and $eV_{\text{bias}} = 0.6\Gamma$. For times larger than a few $\hbar/\Gamma$ (see text), the calculated impurity orientation (plot marks) can be well-fitted to exponentially decaying functions (lines), while higher values of J lead to ever-faster decay- or relaxation times.

5.2. Impact of the Electron-Impurity Interaction on the Charge Current

Before we continue to explore the dependence of the rate W on model parameters such as the interaction strength J and the bias voltage V_{bias}, we proceed by explaining the observed decrease of the current $\langle I \rangle$ for growing

5. Spin-Relaxation by a Charge Current: Exact Results

values of J, which is shown in figure 5.2. Throughout the whole chapter, the following physical situation is simulated. At times $t < 0$, the leads and the dot are uncoupled and in initial state $\hat{\rho}(-\infty)$ [see equation (4.48)], in which the dot contains no electrons and the impurity is in state $|\tau_i =\uparrow\rangle$. Exactly at $t = 0$ the tunnel coupling is switched on *instantaneously* and remains constant for all later times:

$$\Gamma(t) = \Gamma\theta(t). \tag{5.4}$$

For $t > 0$ and bias voltages $V_{\text{bias}} \neq 0$, the current starts to flow and the impurity polarisation decays.[2] Both reach some stationary value at times, which depend on the system parameters. As can be seen in figure 5.1, the time t^{ST}, at which the impurity reaches its stationary value, strongly depends on J and is, for the interaction strengths considered here, *at least* of the order of $\sim 100\hbar/\Gamma$. The current, on the other hand, shows a very fast relaxation behaviour, which is caused by the strong coupling between the leads and the dot. For the parameters considered in this work, it is in fact so fast that we can only estimate the *upper* limit for reaching stationarity being $t^{\text{ST}} \lesssim \hbar/\Gamma$. This is because our particular implementation of the ISPI scheme does not allow to calculate the dynamics for very short propagation times (see section 4.2). Since we are more interested in the transient dynamics of the impurity spin, however, this is not an issue for the further considerations.

Both the grey dots and red crosses in figure 5.2 are obtained by ISPI, while for the latter case only the longitudinal electron-impurity interaction is taken into account. This is done by restricting the path integral to impurity paths with zero number of flip-flops ($m_{\max} = 0$). The solid grey and dashed red lines are fits to a polynomial function $a + bJ^p$, with $p \approx 2.4$ and $p \approx 2.1$, respectively. We performed a simple (lower limit) error estimate

---
[2] The spin-relaxation also occurs at zero bias voltage.

(blue shadings and dotted lines), which is solely based on the statistical (numerical) deviations and how they are propagated by the extrapolation procedures. It is a conservative estimate, since we (i) do not account for further numerical errors that manifest in small imaginary parts of the numerically derived observables and (ii) we do not know or assess the size of possible unaccounted systematic errors in the extrapolation to infinite memory time. Nevertheless, the depicted estimate should provide a reasonable margin of error.

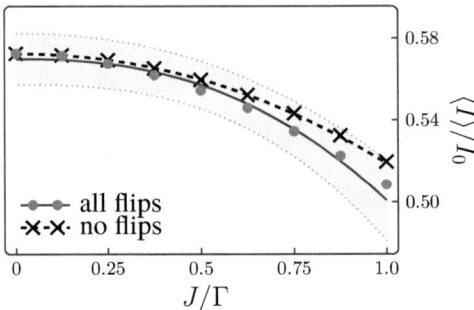

Figure 5.2.: Charge current $\langle I \rangle$ in units of $I_0 = e\Gamma/h$ against J [other parameters as in figure 5.1]. For increasing interaction, the current decreases by about 15% in the range $0 \leq J \leq \Gamma$, showing a nearly quadratic behaviour. The grey solid line corresponds to a fit polynomial $\propto J^p$ with $p \approx 2.4$, while the dotted lines denote a conservative estimate of the error bounds (see text). The dashed line is a polynomial fit ($\propto J^p$) of the current for a purely longitudinal interaction (crosses), when no flip-flop processes are possible and the impurity can be considered as effective magnetic field. It also declines nearly quadratically with $p \approx 2.1$, which suggests that the current is mainly affected by the longitudinal part of the electron-impurity interaction. This is probably due to the relatively high temperature and, consequently, a short coherence time, which strongly limits the influence of coherent dynamics. The growing deviations between the dots and crosses, however, may be an indication of the increasing importance of correlated flip-flops with larger J.

5. Spin-Relaxation by a Charge Current: Exact Results

5.2.1. The Landauer-Büttiker Current

We can draw two main conclusions from figure 5.2. First, the current decreases (nearly) quadratically with J and second, this decrease is mainly caused by the energy shift of the single-electron levels due to the longitudinal part of the electron-impurity interaction, since both the grey solid line and the red dashed curve agree rather well. Whether both curves coincide within the error margin or the observed deviations between the grey and red data points, which grow with increasing J, are a signature of correlated flip-flop dynamics is, at this point, inconclusive. What can be stated clearly, though, is that the flip-flop term $\hat{H}_{\text{int}}^{\perp}$ has a much smaller influence on the charge current at this rather large temperature (incoherent regime) than the longitudinal part of the interaction. To understand this behaviour, it is worthwhile to compare the ISPI results with the Landauer-Büttiker (LB) current $\langle I \rangle_{\text{LB}}$ (see [202]), which is in general given by

$$\langle I \rangle_{\text{LB}} = \frac{e\Gamma}{2h} \sum_\sigma \int_{-\infty}^{\infty} d\omega \, [f_{\text{L}}^+(\omega) - f_{\text{R}}^+(\omega)] \, \Im[\overbrace{(G_\sigma)_\omega^{-+}}^{(G_\sigma)_\omega^r} - (G_\sigma)_\omega^{+-}], \quad (5.5)$$

where G_σ is the (full) Green's matrix for spin σ and the superscript r indicates the retarded component. For the dot with the full Coulomb and electron-impurity interaction, however, we do not possess a closed expression of G_σ. Nevertheless, a non-equilibrium current can still be obtained with the help of this formula, as long as the Green's function that is plugged in for G_σ incorporates at least the interaction with the leads. Hence, if we were interested in the current through the "quasi-free" dot with *switched-off electron-impurity interaction* neglecting any *correlation effects due to the Coulomb interaction*, we could calculate it by plugging-in the free retarded Green's function

5. Spin-Relaxation by a Charge Current: Exact Results

$$(G_{0,\sigma}^{\text{el}})_\omega^r = \lim_{\delta\omega \to 0} \int_{\omega-\delta\omega}^{\omega+\delta\omega} \frac{d\omega'}{2\pi} [(G_{0,\sigma}^{\text{el}})_{\omega,\omega'}^{-+} - (G_{0,\sigma}^{\text{el}})_{\omega,\omega'}^{+-}]$$
$$= \frac{2i\Gamma/\hbar}{(\omega - \omega_\sigma^U)^2 + (\Gamma/\hbar)^2},$$
(5.6)

which we get with the help of equation (G.14). Actually, since $\omega_\sigma^U = (\epsilon_\sigma + U/2)/\hbar$ contains the Coulomb energy U, this Green's function is not really free of interaction but rather it implements the (quasi-"free", classical, or single-particle) part of the Coulomb interaction that can be mapped to shifts of the single-particle energies. With the help of the HS-transformation in section 4.1.4, we identified it with $U/2$, which is independent both of electron spin and the fluctuating HS-fields. In case of the electron-impurity interaction, we can assume that, as a first approximation, the longitudinal term $\hat{H}_{\text{int}}^{\parallel}$ causes a similar effect, viz., an energy shift by $\pm J$. In contrast to the Coulomb interaction, however, this shift depends on the time and (impurity-) path, since its sign is given by the relative alignment of electron- and impurity spin.

In this context, it should be noted that the imaginary part of the retarded Green's function basically gives the local density of states for the dot electrons of different spin. This allows a rather simple physical interpretation of the current formula. The joint density of dot-electron states is just given by a Cauchy-Lorentz (or Breit-Wiegner) function, whose width equals the tunnel coupling strength and whose resonance lies at the single-electron energy ω_σ^U. The (non-interacting) current is then given by the integral over the energy-dependent difference of the left and right lead's occupation multiplied by the density of available dot states at that same energy.

A major question is now, how the density of states is affected by the *longitudinal impurity interaction*. For a fixed orientation of the impurity- and electron spin, $\hat{H}_{\text{int}}^{\parallel}$ causes oppositely directed shifts for each of the single-particle channels that are connected to the empty and doubly occupied dot

5. Spin-Relaxation by a Charge Current: Exact Results

(the green and red dashed lines in figure 5.3). In other words, if the single-level channel for tunnelling into or out of the empty dot is lowered by J for electrons of spin σ, the channel that involves the doubly occupied dot is raised by the same amount (both relative to the channels for $J = 0$). The channels for electrons with spin $-\sigma$ are shifted in the respective opposite directions.

This inspires us to the following mean-field-type ansatz to account for $\hat{H}_{\text{int}}^{\parallel}$. When compared to the case $J = 0$, half of the density of states for each spin is shifted by J, the other by $-J$:

$$(G_{0,\sigma}^{\text{el},J})_\omega^r := \frac{i\Gamma}{\hbar} \sum_{\alpha=\pm 1} \frac{1}{(\omega - \omega_\sigma^U + \alpha J/\hbar)^2 + (\Gamma/\hbar)^2}. \qquad (5.7)$$

In general, we assume that all four Keldysh components of the free Green's function transform accordingly for $J \neq 0$. Besides for the calculation of the LB current, this ansatz will then also be employed below to calculate a rate for a sequential approximation of the flip-flop rates W_τ. With equations (5.5) and (5.7), we can now evaluate a LB current for a quantum dot, whose interactions manifest only in single-particle energy shifts.

In which way an increasing J thus affects the transport conditions is illustrated with the help of the transport scheme in figure 5.3, where the ratios of the depicted energy distances correspond to the parameters used in figure 5.1. As usual, the lightly shaded regions on both sides of the dot symbolise the occupation of the leads, the difference of which is highlighted by a cross hatching (required by the rather small quotient of bias voltage and temperature of 0.6). The black horizontal lines mark the transport channels for $J = 0$. A sweep in the interaction strength J shifts these channels up- or downwards depending on the relative spin orientation between the dot electron and impurity (dashed lines). For $0 \leq J \leq \Gamma$, the sweep direction

5. Spin-Relaxation by a Charge Current: Exact Results

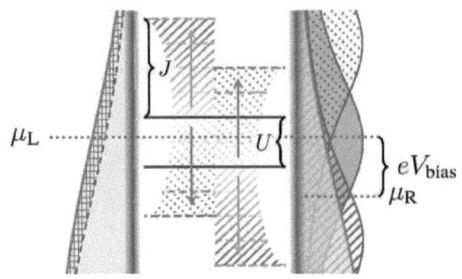

Figure 5.3.: A schematic illustration of how an increasing J between zero and Γ affects the transport environment. The ratio of the energies corresponds to the model parameters from figure 5.1. For $\beta\Gamma = 1$, the occupation of lead states (light shaded areas on left and right) changes slowly compared to the size of the transport window. The bias induced occupation difference is indicated by the cross hatched area on the left, while the black horizontal lines indicate the transport channels for $J = 0$. For positive $J > 0$, an individual electron's energy on the dot is shifted due to the longitudinal part of the impurity interaction. The direction of the shift depends on the mutual orientation of impurity- and electron spin, while its amount is equal to $|J|$. This is indicated by the dashed (virtual) channels and the corresponding arrows. We call them "virtual", since it is not possible to assign a unique (electron) spin to them. This is due the fact that the impurity orientation can change as an effect of the transverse (flip-flop) interaction. For example, if $\tau_z = +1$, the line hatched (dotted) regions correspond to electron spin-down (spin-up). Furthermore, we depicted the local density of states (Lorentz curves to the right) as given by the imaginary part of expression (5.7) for $J = 0$ (grey) and $J = \Gamma$ (hatched).

is indicated by the arrows. Since the impurity spin can change by flip-flop processes, it is not possible to (uniquely) assign an electron spin to these shifted channels, which we therefore call "virtual." Any spin assignment is only time-local and path-dependent. For instance, if the impurity is in the spin-up state at a given time for a given impurity path and the interaction is anti-ferromagnetic, the line hatched (dotted) channels correspond to spin-down (spin-up) electrons. This is also reflected in the local density of dot electron states, which we depict on the r.h.s. of the figure.

For $J = 0$, the density of both spin states is degenerate and its resonance lies between the black transport channels (darkly shaded area), since it is off-

5. Spin-Relaxation by a Charge Current: Exact Results

set by half the Coulomb energy with respect to the single-particle channel. A non-vanishing interaction lifts this degeneracy, shifting the resonance energies for the spin directions upwards and downwards by $\pm J$. Again, a unique assignment of an electron spin to one of the densities is not possible. As long as the charge current for non-polarised leads is considered, however, this is not a problem, as it is only important by *how much* the degeneracy of the channels is lifted due to the impurity interaction. From the figure, we can also read out, why the Landauer Büttiker current has to decrease for growing J. According to equations (5.5) and (5.7), $\langle I \rangle_{\text{LB}}$ is given by the area under product of two functions: (i) the difference between the lead occupations, $f_L^+(\omega) - f_R^+(\omega)$ (depicted by the cross hatched area to left), and (ii) the local density of states in the dot (Lorentz curves on the r.h.s.). It can be seen that the difference in occupation is largest around the Fermi level, where it has the biggest overlap with the darkly shaded density of states for $J = 0$. With increasing J, the density resonances "move away" from the Fermi level, where $f_L^+(\omega) - f_R^+(\omega)$ decreases visibly. As a direct consequence, the current drops.

To quantify the decrease of the LB current in detail, we derive an analytic expression for it, using complex contour integration (see appendix H). This yields

$$\langle I \rangle_{\text{LB}} = \frac{\pi e \Gamma}{2h} \sum_{\sigma,p,\alpha} p \{ \Re f_p^+(\omega_\sigma^U + [\alpha J + i\Gamma]/\hbar) + \sum_{q=\pm 1} q \xi_p^q (\omega_\sigma^U + \alpha J/\hbar) \} \tag{5.8a}$$

with

$$\xi_p^q(\omega) := \xi(\omega - [\mu_p - iq\Gamma]/\hbar) \text{ and } \xi(\omega) := \frac{1}{2\pi} \Im \Psi^{(0)} \left(\frac{1}{2} + \frac{i\beta}{2\pi} \hbar \omega \right), \tag{5.8b}$$

where $\Psi^{(0)}(z)$ is the digamma function and $q = \pm 1$. A comparison of the numerical data points from figure 5.2 and the LB current (solid line, same model parameters) is shown in figure 5.4. It can be seen that the $\langle I \rangle_{\text{LB}}$ behaves similar to the ISPI current with deviations ranging between 2% and 4.2%; it lies almost within the error margin of the numerical results, although the deviations are systematic insofar as the Landauer-Büttiker result is consistently higher than the full-coherent, interacting current. Physically, this result is to be expected, as the quantum fluctuations due to the on-site Coulomb interaction should increase the quantum dot's resistance. We can also see, that all current values show a qualitatively similar functional dependence on the interaction strength J. We therefore conclude, that the observed current drop with increasing J is (mainly) caused by the shift of the dot electron energies due to the longitudinal part of the electron-impurity interaction. It causes a declining joint density of states for dot electrons in the vicinity of the Fermi level. From a engineering point of view this may be good news, since it means that a control current that is used to manipulate the impurity dynamics, is in turn only weakly affected by the dynamics of the localised spin (small back action).

Comparing the ISPI results with full interactions (dots in figure 5.4) with the values obtained for switched-off electron impurity scattering ('no flips', crosses), however, reveals a slowly growing discrepancy between both values. This indicates that, not surprisingly, flip-flop processes become more and more important for increasing interaction strength J. Also, the current with flip-flop scattering is lower than without, meaning that the flip-flop processes increase the resistivity of the dot. For the parameters used here, particularly the rather high temperature of $\beta\Gamma = 1$, this effect is relatively small, both compared to (i) the numerical error of the ISPI values and (ii) the (single-particle) effect of the longitudinal interaction component.

It is worth noticing, that although the solid curve and the grey ISPI points

5. Spin-Relaxation by a Charge Current: Exact Results

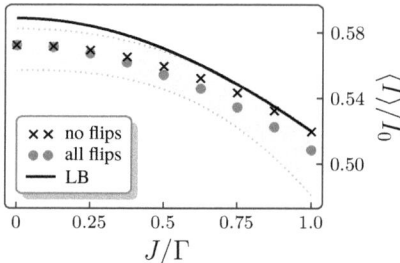

Figure 5.4.: Same as figure 5.2, but now with the Landauer-Büttiker current (LB, green line) for comparison with the ISPI results. Although the LB current lies almost within the error margin (it deviates by at most 4.2% with a minimum of 2%) and exhibits a similar qualitative behaviour, the differences between the grey ISPI values and the LB current are considerable, owing to the presence of the Coulomb- and electron impurity interaction.

have a similar shape, they are far from being congruent; they do not just deviate by a constant value (as if they had been shifted apart vertically) but their differences depend on J. Moreover, while the dots and crosses points are physically consistent (see above), the crosses show *even less agreement* with the LB curve than the dots (with full interaction). Intuitively, the agreement with the LB curve should be *better* without flip-flop scattering (taking into account only the *longitudinal* component of the interaction), as the difference between the crosses and the LB curve is solely due to quantum fluctuations caused by the Coulomb interaction. Therefore, the rather strong qualitative similarities between the LB current and the fully interacting ISPI results may be a kind of "coincidence," caused by a compensation of Coulomb correlation effects by (coherent) flip-flop processes. This does not, however, affect the general validity of our previous reasoning that the current drop is mainly caused by the effect of the impurity interaction on the dot's single-particle energy structure. If anything, it indicates that already in the rather incoherent regime of temperatures around $\beta\Gamma = 1$, correla-

5. Spin-Relaxation by a Charge Current: Exact Results

Figure 5.5.: For parameters $\Phi_D = \Delta = \Delta_{\text{imp}} = 0$, $U = \Gamma$, $\beta\Gamma = 1$, and four different values of the bias voltage, this figure shows how the current $\langle I \rangle$ depends on the interaction strength J. The current decreases with the interaction strength mainly on account of the longitudinal component. The values for different V_{bias} are scaled, so that all fit curves drop by half the figure's height and their start- and end points are one sixth of the height apart. This facilitates the visual comparison of the fit's curvatures, which are quite similar.

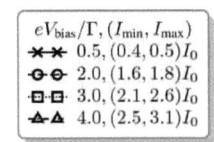

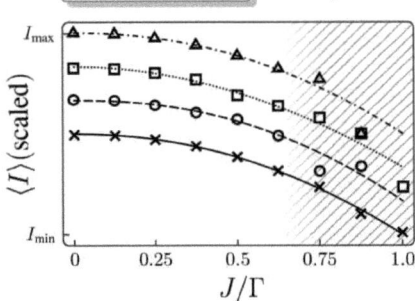

For the present parameter set and $eV_{\text{bias}} \gtrsim \Gamma$, the current value's accuracy drops considerably as soon as J exceeds values of about 0.7Γ. Therefore, we neglected the corresponding data points (in the hatched area) when calculating the fits. This indicates that already for an interaction strength of about $J \sim \Gamma$ and a Coulomb interaction of $U = \Gamma$ relevant correlations persist considerably longer than $\hbar/\Gamma$, making it difficult to obtain reliable simulation results for $K \leq 3$ and reasonable time-steps δ_t. The accuracy may be enhanceable, however, by using higher values of K (memory time) in the simulation.

tion effects due to the two-body interactions play a considerable role for the nonequilibrium current through the dot.

At the end of this section, we shortly discuss how the decrease of the current due to the impurity interaction depends on the bias voltage. Figure 5.5 shows (differently scaled) current curves against J for four values of V_{bias}. All curves comply reasonably well with the explanations given above, as they also decrease "almost quadratically." The relative scaling emphasises this fact, as it allows to compare the curvatures of the plots visually. In the darkly hatched region, however, the accuracy of the numerical results drops significantly. Hence, we did not incorporated them into the calculation of the fit curves (lines). The relatively high numerical error indicates that already for an interaction strength of about $J \sim \Gamma$ *combined with* a Coulomb interaction of $U = \Gamma$, relevant correlations persist considerably longer than $\hbar/\Gamma$. This makes it difficult to obtain reliable simulation results for $K = 2, 3$ and reasonable time-steps δ_t. It may be possible to improve the accuracy, though, by using higher values of K—at the expense of much longer running times.

Note that, except for the current in figures 5.2 and 5.4, we do not present error estimates for all numerical results. On one hand, there are issues considering the comparability of estimates that are obtained by different extrapolation methods for infinite memory time [linear or principle of least dependence, see section 4.2.4]. On the other hand, if the ISPI method is stable in a given parameter range, the errors are mostly of the same order of magnitude, their relative size (in the few to < 10 percent range) comparable with the shown errors of the current and slowly increasing with the interaction strength. If, as for an interaction strength $J \gtrsim 0.7\Gamma$ in figure 5.5, the accuracy of some results is lower-than-average, we point it out explicitly and treat them accordingly.

5.3. First Approximation of Impurity Spin Dynamics: Sequential Flip-Flops

In the next step, we take a closer look at the rate for flip-flop events W or, rather, the inverse relaxation time $\tau_R^{-1} = 2W$ for cases, when equation (5.3) applies. In figure 5.6, we present results of sweeps for J, again with $0 \leq J \leq \Gamma$, but for a Coulomb energy set to $U = \Gamma$ and four different bias voltages. These show a "close to quadratic" behaviour as well, this time growing from zero (no relaxation).

5. Spin-Relaxation by a Charge Current: Exact Results

To investigate the origin of the quadratic dependence of τ_R^{-1} on J, as shown in figure 5.6, we develop a simple theoretical description of the flip-flop rate W, starting from the general equation (2.9) for a time-dependent Keldysh expectation value. Similar to the theory in chapter 3, the subsequent steps taken to arrive at an analytic expression for the rates of sequential flip-flop processes are based on the real-time diagrammatic technique developed by Schoeller and Schön [46]. As a first step, we argue that the electronic part of the dot is already (nearly) stationary on the time scale of the impurity dynamics. This is supported by (i) the observation that the time, on which the charge current assumes its stationary state (upper limit: $t^{ST} \lesssim \hbar/\Gamma$, see section 5.2), is very short when compared to the corresponding time for the impurity spin (lower limit: $t^{ST} \gtrsim 100\hbar/\Gamma$) and also (ii) by the fact that the impurity orientation is well described by a master equation (5.1) with time-independent rates. In itself, the charge current, a spin-independent quantity, does not suffice to indicate the actual stationarity of the electronic system. Since in our model, a time dependence of the current's spin components can only be caused by the time-dependence of the impurity orientation, however, we can estimate that it is negligibly small in the studied parameter range—the orientation of the single spin 1/2 that is the impurity is dissipated only slowly within several tens or even hundreds of $\hbar/\Gamma$.[3]

This said, we can use equation (2.9) to write the probability for finding the impurity spin in state $|-\tau\rangle$ at time Δ_t, under the condition that it was in state $|\tau\rangle$ at time 0, as

$$P(-\tau, \Delta_t | \tau, 0) =: \Pi_{-\tau \leftarrow \tau}^{-\tau \leftarrow \tau}(\Delta_t, 0) = \text{Tr}\{\hat{\rho}_0^\tau \hat{U}(0, \Delta_t) |-\tau\rangle\langle -\tau| \hat{U}(\Delta_t, 0)\}, \quad (5.9)$$

where $\hat{\rho}_0^\tau := |\tau\rangle\langle\tau| \hat{\rho}_{\text{dot}}^{\text{el, st}} \hat{\rho}_{\text{leads}}$ is the (initial) density matrix at time $t = 0$

---
[3]Except for the instantaneous switching at $t = 0$ the Hamiltonian is not explicitly time dependent.

5. Spin-Relaxation by a Charge Current: Exact Results

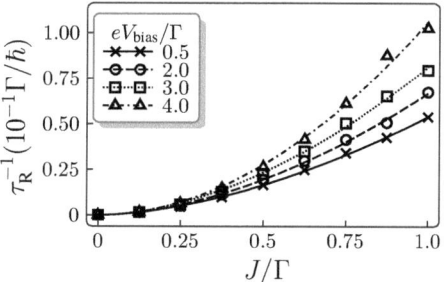

Figure 5.6.: Inverse impurity relaxation time τ_R^{-1} (relaxation rate) versus interaction strength J, for four different values of V_{bias}. The parameters correspond to those from figure 5.5. As in figure 5.1, the solid lines are fits to polynomial functions $\propto J^p$ and, likewise, all resulting values of p are close enough to 2 (between 1.7 and 1.95), to motivate looking into a perturbative description of the relaxation process (see below). The polarisation decays faster with increasing J.

and $\hat{\rho}_{\text{dot}}^{\text{el, st}}$ denotes the stationary state (superscript 'st') of the dot's electronic subsystem. The fourth-rank tensor $\hat{\Pi}$ is the full propagator of the reduced density matrix for the impurity (an analogous operators appears in chapter 3). Now, by inserting the interaction picture version (4.6) of the full time evolution operator $\hat{U}$ with $\hat{H}_1 = \hat{H}_{\text{int}}^\perp$, we obtain the following diagrammatic version of (5.9)

$$P(-\tau, \Delta_t | \tau, 0) = 0 \quad \cdots \quad \Delta_t + 0 \quad \cdots \quad \Delta_t + \ldots \tag{5.10}$$

It requires to sum over all diagrams with an equal number of flip-flop vertices with an *odd* number of vertices on each branch, thus making sure that the impurity spin is actually flipped at $t = \Delta_t$. The diagrammatic symbols are identical to those introduced in section 4.1.1, where the lines stand for

5. Spin-Relaxation by a Charge Current: Exact Results

"non-flipping" time evolution according to Hamiltonian $\hat{H}_0$ and a Keldysh time ordered integration over the vertex positions is implied. The quotes are there to emphasise that the propagation is, in fact, not free at all, as $\hat{H}_0 \equiv \hat{H} - \hat{H}_{\text{int}}^{\perp}$ contains the full Coulomb interaction, the coupling to the leads, and the longitudinal impurity interaction. Each flip-flop vertex contributes a factor $\mp i J/(2\hbar)$, where the sign depends on the contour branch, and two dot electron fields to the diagram. They can thus be written as time integrals over expectation values of dot fields. For example, the first diagram on the r.h.s. of equation (5.10), translates to [203]

$$\begin{array}{c} \tau \\ 0 \\ \tau \end{array} \begin{array}{c} t_1 \\ \boxed{\hat{H}_{\text{int}}^{\perp}} \\ \boxed{\hat{H}_{\text{int}}^{\perp}} \\ t_2 \end{array} \begin{array}{c} -\tau \\ -\tau \end{array} = \frac{J^2}{4\hbar^2} \int_0^{\Delta_t} dt_1 \int_0^{\Delta_t} dt_2 \; {}_0\langle \bar{d}^-_{-\tau}(t_2) d^-_{\tau}(t_2) \bar{d}^+_{\tau}(t_1) d^+_{-\tau}(t_1)\rangle_0.$$

(5.11)

As in section 4.1.8, we would like to employ Wick's theorem to expand the expectation value into a sum of products of Green's matrix elements—the total pairing. This time, however, we have to use the Green's matrix of the model without flip-flops, as described by $\hat{H}_0$ (denoted by the subscript '0'). Since $\hat{H}_0$ contains the two-particle Coulomb interaction a *direct* application of the theorem is actually not possible. Rather, we could again use the Hubbard-Stratonovich transformation to write the expectation value as a path integral (sum) over the HS-field ς:

$$\begin{aligned} {}_0\langle \bar{d}^-_{-\tau}(t_2) d^-_{\tau}(t_2) \bar{d}^+_{\tau}(t_1) d^+_{-\tau}(t_1)\rangle_0 = \\ \lim_{\delta_t \to 0} 2^{-2N} \sum_{\{\varsigma\}} \langle \bar{d}^-_{-\tau}(t_2) d^-_{\tau}(t_2) \bar{d}^+_{\tau}(t_1) d^+_{-\tau}(t_1)\rangle_{\varsigma}, \end{aligned}$$

(5.12)

where $\{\varsigma\}$ denotes a Keldysh path between times t_1 and t_2 and the expectation value with subscript 'ς' has to be taken with respect to action $S_K^{\text{el}}[\{\varsigma\}]$,

5. Spin-Relaxation by a Charge Current: Exact Results

which is obtained from the equations (4.50) and (4.51) by setting $\{\tau\} \equiv 0$. As this action is quadratic in the electron fields, Wick's theorem can be applied to each individual summand on the r.h.s. In principle, we could use this approach to correctly account for the Coulomb interaction. However, evaluation of the path sum would be numerically costly and thwart the advantage that we want to gain from the whole rate equation approach: a first, rather simple and intuitive understanding of the basic impurity dynamics in terms of elementary processes. Hence we make the assumption, that similar to the case of the charge current, we can neglect the effects of fluctuations and account only for the single-particle effects (the energy shift of $U/2$) of the Coulomb interaction:

$$\langle \bar{d}^-_{-\tau}(t_2) d^-_\tau(t_2) \bar{d}^+_\tau(t_1) d^+_{-\tau}(t_1) \rangle_0 \approx \langle \bar{d}^-_{-\tau}(t_2) d^-_\tau(t_2) \bar{d}^+_\tau(t_1) d^+_{-\tau}(t_1) \rangle_\varsigma \big|_{\{\varsigma\}=0}. \quad (5.13)$$

Although this approximation is uncontrolled in the sense that we cannot give a theoretical expression for the error as a function of U, we can estimate that, in the considered parameter range, it will be accurate enough to suit our purposes. We notice that the flip-flop part of the interaction is *directly* connected to the spin relaxation, while the Coulomb fluctuations only influence it *indirectly* via their effect on the electron system, i.e., the charge current. As we consider values of parameter J that are not actually small in a perturbative sense, below we observe discrepancies between the numerically exact and the perturbative dynamics of the order 10%. On the other hand, the (direct) effect of Coulomb fluctuations even on the current turn out to be up to one order of magnitude smaller. Hence, we conclude that as far as the impurity dynamics is concerned, neglecting higher-order perturbation terms in J—a systematical approximation—entails errors at least on order of magnitude larger than those caused by neglecting the Coulomb fluctuations. For our purposes, this approximation is sufficiently accurate.

With this, we can apply Wick's theorem to the r.h.s. of equation (5.13),

5. Spin-Relaxation by a Charge Current: Exact Results

which then equals the total pairing. Every summand of the total pairing can again be represented by a diagram, where each contraction for each spin direction is symbolised by a directed line, connecting two flip-flop vertices. At a creator (annihilator) dot electron field, a contraction line starts (ends). Hence, a vertex that flips the impurity spin from τ to $-\tau$ is connected to both an incoming electron line with spin $\sigma = -\tau$ and an outgoing line with $\sigma = \tau$.

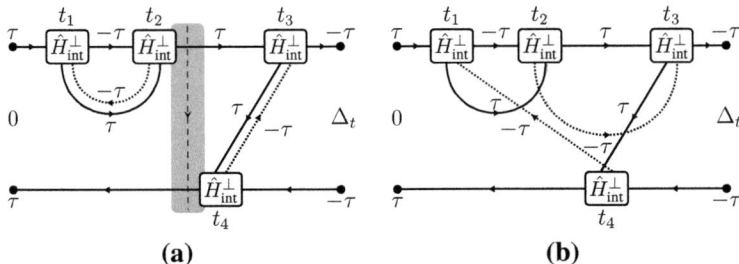

Figure 5.7.: Examples for two flip-flop diagrams of fourth order in $\hat{H}_{\text{int}}^{\perp}$. **(a)** This diagram is reducible, for it can be vertically cut (dashed line), without cutting one of the contraction lines. During the time period between the latest vertex before and the earliest vertex after the cut (grey shaded region), the system propagates "freely," i.e., without any flip-flops (but is subdue to electron tunnelling and the full Coulomb interaction). **(b)** Example of an *irreducible* diagram. At all times between the earliest and latest vertex, a vertical line crosses *at least one* contraction line. In this case, the short propagation lines at the very ends of the diagram are just drawn to carry the state labels of the real-time early and late contour ends. For the sake of the argument, they should be considered as infinitely short.

In the last step, the real-time order of all vertices regardless of the contour branch is fixed for all the diagrams that are constructed in this way. In doing so, we make sure that the integrals, into which the resulting diagrams translate, have continuous functions as integrands; we discussed in section 4.2.3 and illustrated with figure 4.7, that the Green's function jumps discontinu-

5. Spin-Relaxation by a Charge Current: Exact Results

ously at coinciding times $t = t'$.[4] More importantly, the real-time order of the vertices can be decisive when it comes to a property that is particularly important to determine flip-flop rates: a diagram's so-called *reducibility* (see also chapter 3). With respect to this property, all diagrams with fully contracted vertices and fixed real-time order can be divided into two (disjunct) groups. If a diagram can, at some point (between its earliest and latest vertex), be cut vertically without crossing a contraction line it is called reducible, otherwise irreducible. Figure 5.7 presents an exemplary diagram for each of both cases. In the whole grey highlighted region of sub-figure **(a)**, a vertical cut (dashed line) leaves the contraction lines intact. Hence, the diagram is reducible, while the aforesaid grey region marks a period of (flip-flop-) "free" time evolution according to $\hat{H}_0$. The diagram of sub-figure **(b)**, however, can not be vertically cut without crossing a contraction line. Hence, it is irreducible.

As we will see below (and have already seen in chapter 3), the distinction between reducible and irreducible diagrams is vital, as only the irreducible diagrams contribute to a transition rate. The reason for this lies in the fact, that irreducible diagrams are directly connected to and describe *complex, coherent processes*. This is illustrated best with the help of an example. Both diagrams in figure 5.7 describe an effective flip-flop process between times 0 and Δ_t; the system starts in impurity state $|\tau\rangle$ at $t = 0$ and ends in the state with opposite orientation at time Δ_t. But while the process in (a) can be viewed as concatenation of two different incoherent (sequential) processes of second-order (before and after the cut), the indivisible process in (b) is coherent over the whole time period. As a consequence, if we know everything about second-order processes, we do so for processes such as the left in figure 5.7, as well. Only by including coherent higher-order flip-flops

---
[4]This not only applies to the free Green's function, as this property is a consequence of the exchange of order of particle creation and annihilation.

5. Spin-Relaxation by a Charge Current: Exact Results

as described by 5.7(b) we can then further increase our knowledge about the system's flip-flop dynamics.

To fully exploit this conceptional distinction, we apply the *Dyson equation* (see, e.g., [186]) to expression (5.10). By realising, that each of the diagrams contributing to the flip-flop probability are either irreducible or can be decomposed into irreducible sub-processes (connected by periods of flip-less propagation), we can write

$$\hat{\Pi} = \hat{\Pi}_0 + \hat{\Pi}_0 \hat{\Sigma} \hat{\Pi}_0 + \hat{\Pi}_0 \hat{\Sigma} \hat{\Pi}_0 \hat{\Sigma} \hat{\Pi}_0 + \ldots = \hat{\Pi}_0 + \hat{\Pi}_0 \hat{\Sigma} \hat{\Pi} \qquad (5.14)$$

with $(A \cdot B)_{\tau_1 \leftarrow \tau_1}^{\tau'_2 \leftarrow \tau_2}(t',t) = \sum_{\tau,\bar{\tau}} \int_t^{t'} dt'' \, A_{\tau'_1 \leftarrow \tau}^{\tau'_2 \leftarrow \bar{\tau}}(t',t'') \, B_{\tau \leftarrow \tau_1}^{\bar{\tau} \leftarrow \tau_2}(t'',t),$

where $\hat{\Pi}_0$ denotes the (flip-flop-) "free" propagator, operator $\hat{\Sigma}$ is given by the sum of all *irreducible* diagrams, and $A, B = \Pi, \Pi_0, \Sigma$. By plugging the r.h.s. of the Dyson equation into (5.10), we arrive at an identity that expresses the probability for a spin flip at time Δ_t solely by the irreducible kernel $\hat{\Sigma}$ and the probability to find the spin in either of its eigenstates at earlier times $0 < t < \Delta_t$ (for details, see appendix H). Since $\hat{\Pi}_0$ cannot change the impurity spin in our system, the first term on the r.h.s. of equation (5.14) does not contribute to $P(-\tau, \Delta_t|\tau, 0)$. Since we are not interested in the actual flip-flop probability but rather the flip-flop rate, i.e., the flip probability *per time*, we derive with respect to Δ_t and get (appendix H)

$$\frac{dP}{d\Delta_t}(-\tau, \Delta_t|\tau, 0) = \sum_{\tau'} \int_0^{\Delta_t} dt \, \Sigma_{-\tau \leftarrow \tau'}^{-\tau \leftarrow \tau'}(\Delta_t, t) P(\tau', t|\tau, 0). \qquad (5.15)$$

It can be seen from this equation, that as soon as flip-flop *rates* are considered, the irreducible diagrams (as represented by $\hat{\Sigma}$) play the essential role. Note that this formally exact kinetic equation is not Markovian—the probability at time Δ_t depends on all intermediate probabilities to find the

5. Spin-Relaxation by a Charge Current: Exact Results

impurity in one of its eigenstates.

Thus, to make it usable for us, we now exploit the assumptions that the electronic system is stationary and the rates are time-independent. They are embedded in the rate equations (5.2), which predict that the impurity spin reaches a stationary state, as well. For the cases studied this chapter, it is characterised by a vanishing polarisation [see equation (5.3)]. If the flip-flop rates are stationary for all times, however, they have to have the same constant value also in the stationary regime, in which equation (5.15) still applies. Hence, we can set $P(\tau', t|\tau, 0) \equiv P_{\tau'}^{st} = 1/2$ to pull the probabilities out of the integral on the r.h.s. of equation (5.15), while the l.h.s. vanishes. Also, we have to set the time Δ_t to infinity, as the equation has to be fulfilled for arbitrary long positive propagation times, if the system is already stationary at $t = 0$. This is again an example of the idealisation of an infinite limit. Since the integrand, basically a polynomial of Green's matrix elements, decays exponentially with time, the integral converges fast for growing Δ_t and, hence, setting it to infinity implements the assumption, that the time scale for coherent flip-flops is limited only by the system's memory time. The flip-flop rates can then be identified with the integral

$$W_{-\tau} = -\int_0^\infty dt\, \Sigma_{-\tau \leftarrow \tau}^{-\tau \leftarrow \tau}(t, 0) \tag{5.16}$$

over all irreducible diagrams with the appropriate real-time late and early state entries.

Due to the infinite number of diagrams, of which $\hat{\Sigma}$ is composed, we cannot calculate it exactly. However, in view of the dependence of the relaxation time on the interaction strength J—as figure 5.6 shows, it is nearly quadratic—, we can assume that of all diagrams contributing to equation (5.16), the second-order terms are dominant in this rather incoherent regime. As can be seen from equation (5.11), those diagrams also scale

5. Spin-Relaxation by a Charge Current: Exact Results

with J^2. By restricting the irreducible kernel to only the lowest-order diagrams, we obtain the following approximate flip-flop rates:

$$W^{(2)}_{-\tau} = -\int_0^\infty dt \left\{ \begin{array}{c} \text{[diagram 1]} \end{array} + \begin{array}{c} \text{[diagram 2]} \end{array} \right\}, \quad (5.17)$$

where the superscript integer denotes the vertex-order. These are the simplest possible irreducible diagrams and describe incoherent, sequential flip-flops in the system. Conversely, a quadratic dependence of the inverse relaxation time on J suggests, that the impurity dynamics is dominated by sequential flip-flops.

As we explained above, the exact Green's function for the system with Coulomb interaction is not known and we only calculate an approximation of this rate as given by the generalisation of equation (5.13) to arbitrary operator polynomials. This is equivalent to using the non-interacting Green's function $(G^{\text{el}}_{0,\sigma})_{\omega,\omega'}$ from equation (G.14), which contains, besides the full coupling to the leads, an energy shift of $U/2$ (obtained by the Hubbard-Stratonovich transformation). Finally, analogous to equation (5.7), we transform to the phenomenological, mean-field-type Green's function $(G^{\text{el},J}_{0,\sigma})_\omega$ to account for the longitudinal part of the impurity interaction, as well, and get (see appendix H)

$$\tau_R^{-1} \approx \frac{J^2 \Gamma^2}{16\pi \hbar^4} \sum_{\alpha,\alpha'=\pm} \int_{-\infty}^\infty d\omega \times \cdots$$

$$\times \frac{[f_L^+(\omega) + f_R^+(\omega)][f_L^-(\omega) + f_R^-(\omega)]}{[(\omega - \omega^U_\uparrow + \alpha J/\hbar)^2 + (\Gamma/\hbar)^2][(\omega - \omega^U_\downarrow + \alpha' J/\hbar)^2 + (\Gamma/\hbar)^2]},$$
(5.18)

where we used that $\tau_R^{-1} = W_+ + W_-$. We content ourselves with the inte-

5. Spin-Relaxation by a Charge Current: Exact Results

gral expression for the inverse rate here, as it is not only easily evaluated numerically, but—similar to the integral LB current—it reveals the physical structure and allows for the intuitive interpretation of the processes behind sequential flip-flops. In the numerator of the integrand, we have the sum of all four possible ways to multiply one of the lead's occupations with another or the same lead's probability to find an empty state at some energy. Each of these four combinations is then multiplied by the Lorentz density for the two different spin states each shifted by $\pm J$. This suggests the following interpretation. A sequential flip-flop process consists of three "atomic" components: the actual flip-flop and two tunnelling processes of single electrons with opposite spin (not necessarily in that order). Since they evolve coherently, these components form an effective spin-flip process $|\chi, \tau\rangle \rightarrow |\chi, -\tau\rangle$, where $\chi \in \{0, \sigma, d\}$ and the underlying flip-flop nature is masked by the tunnelling electrons.[5] To every choice of α and α' in the sum (5.18), we can assign certain such effective flip processes.

We illustrate this by an example. Suppose the impurity is in state $|\tau =\uparrow\rangle$. From the diagrams in equation (5.17), we can then read out that, during the process, a spin-$\downarrow$ electron tunnels *onto* the dot via lead p [red contraction line, factor f_p^+ in the numerator of equation (5.18)], while a spin-$\uparrow$ electron tunnels *out* via lead p' [blue line, factor $f_{p'}^-$]. Thus, by fixing τ, we already know all three atomic processes, but neither in which order they occur nor which states they connect. This can be fixed, however, as soon as the electronic part of the dot state is specified. If, for example, the dot is in state $|0, \uparrow\rangle$, the only possible path in state space to arrange a flip-flop, an incoming spin-$\uparrow$ electron, and an outgoing spin-$\downarrow$ electron is the following:

(5.F1) The spin-$\downarrow$ electron tunnels onto the dot: $|0, \uparrow\rangle \rightarrow |\downarrow, \uparrow\rangle$.

---

[5]The dot electron state does not change.

5. Spin-Relaxation by a Charge Current: Exact Results

(5.F2) The actual flip-flop occurs: $|\downarrow,\uparrow\rangle \to |\uparrow,\downarrow\rangle$.

(5.F3) The spin-$\uparrow$ electron leaves the dot: $|\uparrow,\downarrow\rangle \to |0,\downarrow\rangle$.

The relevant density of states for process (5.F1) is the Lorentzian with a resonance at the single-particle energy for spin-$\downarrow$ electrons shifted by $-J$ due to the longitudinal impurity interaction: $\omega_\downarrow^U - J$. Accordingly, the density for process (5.F3) is peaked around $\omega_\uparrow^U - J$. Hence the flip process $|0,\uparrow\rangle \to |0,\downarrow\rangle$ is represented by the summand with $\alpha = \alpha' = 1$ in integral (5.18). Analogously, we can proceed when the dot is in the doubly occupied ($\alpha = \alpha' = -1$) or one of the single-electron states ($\alpha = \sigma = -\alpha'$). In case of the latter ($\chi = \sigma$), the same integrand represents two possible paths for the impurity spin flip, as the system can propagate through either the empty or doubly occupied dot.

All these possible ways, in which the effective flip can take place, add up to the sum in expression (5.18), and conversely, each of the possible ways to choose a combination of Fermi functions and densities corresponds to a particular flip-flop process. The probability density for a spin-σ electron of a certain frequency ω to tunnel into the dot via lead p is given by the product of $f_p^+(\omega)$, the occupation, times the density of states $\propto 1/[(\omega - \omega_\sigma^U \pm J)^2 + (\Gamma/\hbar)^2]$. An analogous probability can be assigned to the second tunnelling out of the dot. Thus, the (sequential) inverse relaxation time (5.18) is basically given by the total probability density of the two necessary tunnelling processes to occur at a certain energy, integrated over all energies and summed over all possible paths in the state space for every transition $|\chi,\tau\rangle \to |\chi,-\tau\rangle$.

As a first step, we use the sequential flip-flop approximation to study a "region" of the deep-quantum regime, where we can expect it to yield the best

5. Spin-Relaxation by a Charge Current: Exact Results

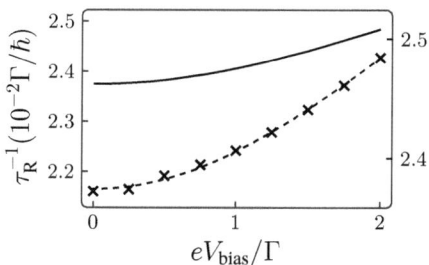

Figure 5.8.: Comparison of the numerically exact [ISPI, crosses] and the sequential [solid line] inverse relaxation time as they depend on the bias voltage. The parameters are $J = \Gamma/2$, $\beta\Gamma = 1$, and $\Phi_D = \Delta = \Delta_{\text{imp}} = U = 0$. Quantitatively, the sequential and ISPI relaxation times differ by up to 10%. This is not surprising, since J is not a small parameter in the system and we can not expect that all rates of higher-order in J are negligible. Also, some of the deviations may result from using the phenomenological Green's function $(G_{0,\sigma}^{\text{el},J})_\omega$. It is notable, however, that both curves exhibit quite similar features on a qualitative level. This is illustrated by the rescaled curve (scale on the r.h.s.) for the sequential decay time (dashed line), which agrees well with the ISPI data points. For small voltages, the relaxation speed increases slowly, similar to some power law, while it grows faster, more or less linearly, for $eV_{\text{bias}} \gtrsim 1.25\Gamma$. Therefore, we can assume that the intuitive picture of sequential flip-flops is also valid to explain the qualitative behaviour of these (particular) ISPI values, which describe the rather incoherent regime of large T (see text).

possible agreement with the numerically exact ISPI results:[6] the case of relatively small electron-impurity interaction ($J = \Gamma/2$), *vanishing Coulomb repulsion* $U = 0$ and rather high temperature $\beta\Gamma = 1$. In this configuration, J is the smallest parameter (though it is not small in a perturbative sense) and the temperature suppresses long time correlations. This is done for two reasons: (i) to test the general validity/adequacy of the sequential approximation and, by providing an explanation for the quadratic dependence of the relaxation rate on J, (ii) gain some first (limited) insight into the physical processes that determine the impurity dynamics (see the beginning of this section). Of course, since the sequential rate (5.18) contains *no correla-*

---

[6]"Best possible agreement" should not, however, be confused with "good agreement."

5. Spin-Relaxation by a Charge Current: Exact Results

tion effects whatsoever we only expect to see (some) agreement to the ISPI rate in cases, where the effect of higher order correlations is negligible or very small (as in the rest of this section). Conversely, with the completely incoherent sequential approximation, we posses a tool that may be useful to spot signatures of correlation effects in the impurity dynamics (see next section).

In figure 5.8 is shown, how the ISPI result (crosses) compares to the sequential relaxation time (solid line). Although the latter is of the right order of magnitude, it is systematically larger than the exact value, differing considerably by up to $\sim 10\%$. Since J is not a small energy parameter of the system, we can presume that the deviations can mostly be attributed to coherent higher-order flip-flop processes, which are neglected in equation (5.18). Another likely source of systematic errors is the usage of the phenomenological, free Green's function $(G_{0,\sigma}^{\text{el},J})_\omega$ for the derivation of the rates. Nevertheless, when it comes to qualitative features, both results agree rather well. This is shown with the help of the rescaled curve for the sequential rates (dashed line). Both the sequential and the exact τ_R^{-1} grow monotonically from their finite value at zero bias voltage. For small voltages, this increase is similar to a power-law (nearly quadratic behaviour), for voltages larger than about $eV_{\text{bias}} \gtrsim 1.25\Gamma$, the inverse relaxation time rises more or less linearly.

Within the intuitive picture of sequential flip-flops we developed above, these features become understandable. In the integral form of the decay time (5.18), the growing bias voltage affects the integrand's numerator via the Fermi functions. Particularly those two of four possible products of Fermi functions with differing lead index. As we explained above, the possibility for sequential flip-flops in the system is directly connected to the number of available lead electrons and holes for the two tunnelling particles, involved in the process. This number, in turn, scales with the area un-

5. Spin-Relaxation by a Charge Current: Exact Results

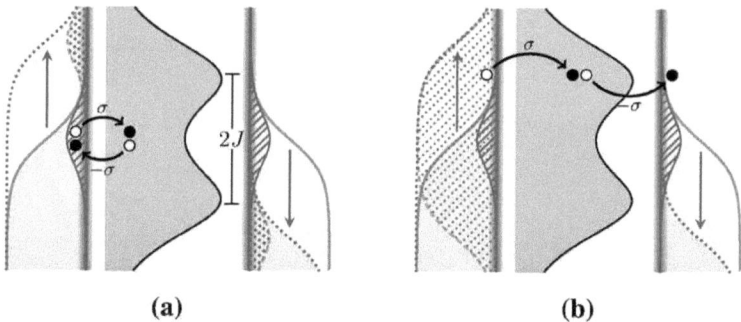

(a)   (b)

Figure 5.9.: Illustration how the sequential relaxation time depends on the bias voltage in a system without Coulomb interaction. The total squared density of states [sum over all α and β in equation (5.18)] in the dot is depicted by the darkly shaded curve between the tunnelling barriers, while the line hatched and dotted areas correspond to the number of lead particles available for the two-electron tunnelling processes that are involved in a sequential flip-flop. **(a)** If both electrons tunnel through the same barrier (thick arrows), the available states are confined to small regions around the Fermi levels, which are shifted by a change of V_{bias} (thin arrows) while their size and shape stay unaltered. Thus, unless the bias voltage is not considerably larger than $2J$, the flip-flops caused by these processes do not vary largely. **(b)** In case that the electrons tunnel through different barriers (thick arrows), the voltage sweep affects the number of available lead particles drastically, while now the centres of the corresponding (hatched and dotted) areas are unaltered. The number of states for tunnelling through the dot from L to R (shown on the l.h.s.) grows to a multiple of its zero-bias value (outlined with dashed line). At the same time, the area for tunnelling in the opposite direction (on the r.h.s) shrinks to zero. For small voltages, the relaxation grows only slowly as the increase of right moving electrons is almost compensated by the declining number of left movers. At some point, the number of right moving electrons grows fast and more or less linearly, while tunnelling to the left is practically non-existent. Eventually, for even higher voltages, the relaxation saturates.

der the product functions $f_{p_1}^+ f_{p_2}^-$ and their degree of overlap with the Lorentz single-particle densities at resonance points $\pm J$. For $p_1 = p_2$, the area under the Fermi product is unaltered by an increase of V_{bias} but merely shifted towards the single-particle resonances for $0 \leq eV_{bias} \leq 2J$ and away from them for even higher voltages. In the end, this leads to a first small increase of the relaxation due to these terms when eV_{bias} approaches $2J$ followed by a slow decrease for eV_{bias} considerably bigger than $2J$. Due to the rather large width Γ (HWHM) of the local densities, though, these variations are smeared out over an accordingly large voltage range and the relaxation can thus be regarded as approximately constant in the bias range considered here. This is shown in figure 5.9(a).

The area under the products $f_p^+ f_{-p}^-$, on the other hand, changes drastically with the bias voltage. The number of available electrons in the leads for tunnelling through the dot *against the voltage drop* rapidly goes to zero with rising V_{bias}, while the corresponding number of lead electrons to tunnel *with the current direction* grows fast and, eventually, linearly. Figure 5.9(b) illustrates this situation graphically. It can be seen that for small bias voltages the increase in relaxation speed due to tunnelling with the current flow is nearly compensated for by the decline in relaxation caused by electrons that go against the current. This accounts for the rather slow growth of τ_R^{-1} for $\epsilon V_{bias} \lesssim 1.25\Gamma$ in figure 5.8. When the voltage grows further, the almost linear increase of the area under $f_L^+ f_R^-$ dominates the bias dependence of the relaxation—flip-flops due to tunnelling against the current become completely negligible. Eventually, for voltages that considerably exceed $2J$, the relaxation rate saturates (not shown).

5.4. Influence of Temperature, Bias Voltage, and Coulomb Interaction

We complete this chapter by studying how the relaxation rate and current are influenced by a finite Coulomb interaction for $J = \Gamma$ at different temper-

5. Spin-Relaxation by a Charge Current: Exact Results

atures. In view of our derivation of the sequential rates τ_R^{-1} from equation (5.18), specifically of the assumptions involved, we expect to find stronger deviations from the ISPI values than for $U = 0$ and $J = \Gamma/2$, the case shown above. First of all, a larger J generally entails an ever increasing importance of higher-order, coherent (non-sequential) processes. Since the temperature is closely related to the lead-induced coherence time τ_c of the system, we expect an increasing prevalence of non-sequential flip-flops for declining T. Also, a faster relaxation leads to a larger, time-dependent imbalance between the spin-components of the charge current and violates the assumption of a stationary electronic- and dynamical state. In other words, an overall stronger relaxation may lead to (slightly) time-dependent relaxation rates. Other obvious simplifications that are "less justified" for stronger interactions regard the free Green's function (5.7): neither does it contain the non-trivial effects of the Coulomb interaction (as included by the sum over fluctuating HS-fields) nor is the mean-field-type way, in which we introduced the longitudinal energy shift J into $(G_{0,\sigma}^{\text{el},J})_\omega$, by any means exact. In general, what we state for the relaxation time, is also valid for the current. As the Coulomb- and impurity interaction increase, the differences between the numerical ISPI results and the LB current (5.8) will, by trend, grow. Particularly at lower temperatures, we expect that the charge current is not independent of the flip-flop interaction $\hat{H}_{\text{int}}^\perp$ any more. Hence, although the LB current is still a more valuable approximation for stronger interactions, compared to the sequential flip-flop theory, we should expect to see considerable deficiencies of the mean-field-type LB current to adequately describe the physics at hand. Nevertheless, comparing the ISPI results with the incoherent LB values can help to identify the signatures of correlation effects.

In figure 5.10, results of a voltage sweep for $0 \leq eV_{\text{bias}} \leq 4\Gamma$, $J = \Gamma$, $U = \Gamma/2$ and temperatures $\beta\Gamma = 1, 2$, and 5 are given. The ISPI values

5. Spin-Relaxation by a Charge Current: Exact Results

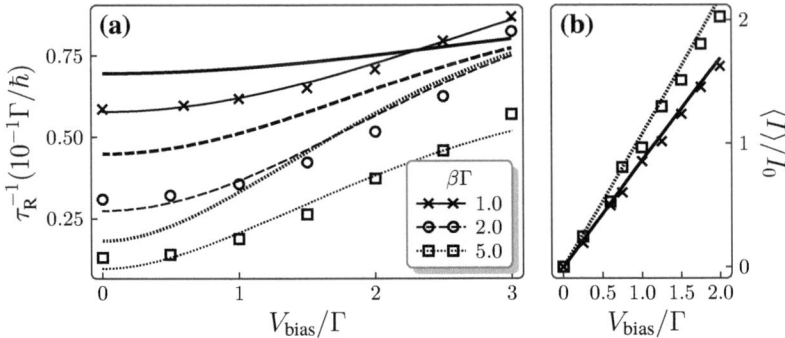

Figure 5.10.: Inverse relaxation time and current versus bias voltage for three (current: two) different temperatures and $U = \Gamma/2$, $J = \Gamma$ (parameters not explicitly given are zero). **(a)** Shown are the ISPI values (plot marks) and the sequential results, where the thick curves give their actual values and the thin lines (of same texture) are rescaled individually for each temperature to minimise the quadratic deviations to the ISPI points. When compared to the system without Coulomb- and with smaller impurity interaction (figure 5.8), we see larger relative deviations between the ISPI results and the sequential flip-flop approximation. The qualitative differences are also stronger: The indication of a saturation behaviour featured by the sequential relaxation curves for bias voltages $\gtrsim 2.5\Gamma$, particularly for $\beta\Gamma = 2, 5$, cannot be read out from the ISPI data. Reasons for the observed discrepancies are: (i) increasing relevance of coherent flip-flop processes of higher-order in $\hat{H}_{\text{int}}^{\perp}$ for growing J and smaller T (longer memory time), (ii) the considerable size of the Coulomb interaction, that we could not fully account for in the derivation of the sequential flip rates, and, partly as a consequence of that, (iii) the usage of an only approximate (free) Green's function for calculating the sequential relaxation. **(b)** The LB current (lines) and the ISPI current (plot marks) coincide reasonably, showing a monotonous, almost structureless behaviour. As in figure 5.4, the approximate LB current based on Green's function (5.7) is larger the the numerical results.

of the inverse relaxation time in **(a)** are given by the plot marks, while the thick lines are the corresponding sequential rates. For each temperature, the thin lines are rescaled versions of the thick curves of the same texture that minimise the squared deviations to the numerical data (as in figure 5.8). In accordance with the simple dynamical picture from figure (5.9), the sequential rates exhibit the typical features (power-law growth, followed by a (quasi-)linear behaviour, which finally saturates), which are the more pronounced the lower the temperature is. As expected, the ISPI data points deviate considerably from this lowest-order approximation. Also, the degree of both the quantitative differences and the deviations in the qualitative behaviour increase with lower temperatures. The ISPI current in sub-figure (b) is compared to the LB current (lines). While the latter is consistently larger than the numerical result (similar to figure 5.4), both coincide quite well and grow monotonically from zero.

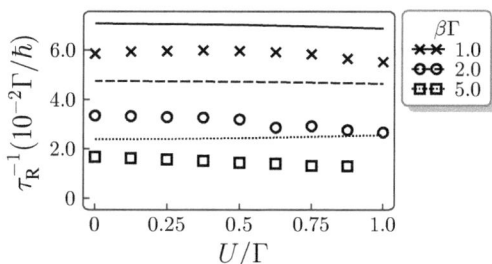

Figure 5.11.: Inverse relaxation time against the Coulomb interaction strength U for three different temperatures and $J = \Gamma$, $eV_{\text{bias}} = 0.6\Gamma$ (all other parameters as in the figures above). For τ_R^{-1}, the absolute differences between the simulation results (plot marks) and the sequential approximation (lines) are comparable for all three temperatures. The relative deviations for smaller T ($\beta\Gamma = 5$), however, are considerably large, reaching a magnitude of up to 100% (see text). Where the low-temperature relaxation rate drops, the sequential theory predicts a rising rate (square marks and dotted line).

Figure 5.11 illustrates the effect of a change in Coulomb energy between $0 \leq U \leq \Gamma$ on the relaxation rate. As before, the solid lines give the se-

5. Spin-Relaxation by a Charge Current: Exact Results

quential flip-flop rates. Quantitatively, the sequential theory agrees with the numerics, insofar as it predicts the relaxation to change rather weakly with U. Particularly for the lower temperatures, however, the relative deviations from the ISPI results are strong: the maximum relative difference is as high as almost 100% (for $\beta\Gamma = 5$, $U = 0.875\Gamma$ and relative to the ISPI value) and while the ISPI rates decline for $\beta\Gamma = 5$, the sequential rates grow.

For all the previous considerations and although it is based on an effective single-particle theory, the LB current proved to be a useful approximation of the nonequilibrium electron (charge) dynamics. As we focussed on the relaxation rate of the impurity spin, rather high temperatures and small (or even vanishing) interactions, this is not surprising. In the regime of intermediate interactions (in particular for stronger Coulomb repulsion) and low temperatures, however, the numerical results will eventually show significant discrepancies to the LB theory. As the latter is purely incoherent, such deviations directly hint at the presence and relevance of coherent dynamics. Therefore, the remainder of this chapter is devoted to a study of certain correlations effects that appear in the nonequilibrium current for intermediate interaction strength $U, J \sim \Gamma$. In figure 5.12, four different current curves at two different temperatures are shown—one for each possibility to either have (i) only mean field dynamics (LB), (ii) the full Coulomb interaction ("no flip") without and flip-flop processes, (iii) flip-flop dynamics without many-body Coulomb correlations, or (iv) the fully interacting dot ("full int."). The Coulomb energy is varied over the interval $0 \leq U \leq \Gamma$, while the electron-impurity interaction and the bias voltage are set to $J = \Gamma$ and $V_{\text{bias}} = 2\Gamma$, respectively. The situation "no U" is implemented by setting Φ_D to $U/2$ (the single-electron part of the Coulomb energy) and the HS-parameter λ to zero. Sub-figure (a) and (b) show the results for $\beta\Gamma = 1$ and $\beta\Gamma = 5$. respectively. Only for the "single-interaction" currents, we present margin of errors, where the shaded area belongs to "no U" and the

5. Spin-Relaxation by a Charge Current: Exact Results

values of "no flip" are tagged with usual error bars (see below).

As long as the Coulomb interaction is small, all current values in both sub-figures lie close. The only physical difference between the currents at this point regards the inclusion or exclusion of the *transversal component* of the electron-impurity interaction (flip-flop processes). Hence the rather good agreement of the $U = 0$ values suggests, that even for lower temperatures flip-flop processes (alone) only weakly affect the current. Particularly in figure 5.12(a), the congruence of the "no U" current with the LB result is remarkable. For $\beta\Gamma = 5$, on the other hand, these curves differ by a slowly increasing that has doubled at the upper end of the considered U interval compared to $U = 0$. Although the error of the ISPI values allows no conclusive statement, it seems that for decreasing temperatures, the flip-flop processes start to influence the current, resulting in an additional resistivity. In case of the "no flip" current, the impurity is fixed and the physical situation is equivalent to a Coulomb-interacting SLQD in a (effective) magnetic field. Both curves in **(a)** and **(b)** show a very similar dependence on U, featuring a local maximum of at around $U = \Gamma/2$; for the lower temperature, the relative height of the broad current peak is twice as big as for $\beta\Gamma = 1$. This is an interesting effect, as we have an increasing degree of correlation effects and still the dot's resistivity decreases (at least for $U \lesssim \Gamma/2$). We suppose that the Coulomb correlations yield to an effective joint density of states in the dot, that differs considerably from the simple mean-field-type that was used to obtain the LB results.

The main reason, why no margin of confidence is indicated for the fully interacting case, regards the comparability of the error data. Calculating the fully interacting current is a very time-consuming task (and more unstable than the time-local impurity polarisation), which is why, at this point, the extrapolation to $\delta_t \to 0$ only involves two time steps. They correspond

5. Spin-Relaxation by a Charge Current: Exact Results

to $K = 3, 4$ for a given coherence time, as the relevant correlations in this parameter regime live too long to yield adequate values for $K = 2$. This does not, however, render these values completely unreliable: we still see a compelling linear behaviour of the $1/\tau_c$ extrapolation (with errors in the range of 1% based on the sample standard deviation) and can compare them to data points, for which more time steps can be calculated. For $U = 0$, for example, we can calculate a value with the same extrapolation parameters as the squares (the stars at $U = 0$) and compare them to the "fully extrapolated" value. The stars in both sub-figures deviate from the exact value by 3.3% and 2.5% and the figures are scaled to show only a small window of current variations of about 8% and 16% around the $U = 0$ value for $\beta\Gamma = 1$ and $\beta\Gamma = 5$, respectively. Hence, we want to stress, that in view of the figure's scaling it can still be said that all current values agree quite well. If the sub-figures would show the current starting from $\langle I \rangle = 0$, the error bars were barely visible and, accordingly, all the data points lay close as in figure 5.10(b). Furthermore, the subsequent squares up to about $U = \Gamma/2$ lie on a increasing slope, which is quite similar to the behaviour shown by the "no flip" crosses. At least for Coulomb interactions not exceeding $U = \Gamma/2$ it seems, that the lack of data points for the extrapolation $\delta_t \to 0$ may merely results in a few-percent shift to lower current values, while the relative positioning (the shape of the curve) is more or less unaffected.

The situation changes for $U \gtrsim \Gamma/2$, where the data points start to show a high variability and the numerics are less stable. Whether the strong variations that can be seen in the case of stronger Coulomb interactions, are the signatures of a physical (correlation) effect or an numerical artefact remains unclear for now. In the future, it may be possible to enhance the accuracy and stability of the simulation, though, by using higher values of K. If our previous interpretations are correct, we can draw an interesting conclusion from the rising current slopes in **(a)** and **(b)**. As the relative increase of the

5. Spin-Relaxation by a Charge Current: Exact Results

Figure 5.12.: This figure compares (i) the LB current (dashed), (ii) the Coulomb interacting current without any flip-flop scattering ("no flips"), (iii) the current without Coulomb scattering but full impurity interaction ("no U", see text), and (iv) the fully interacting current ("full int.") in their dependence on the Coulomb interaction U for $\beta\Gamma = 1$ [sub-figure **(a)**] and $\beta\Gamma = 5$ [sub-figure **(b)**]. The other (non-zero) parameters are $J = \Gamma$ and $V_{\text{bias}} = 2\Gamma$. The shaded areas indicate the error margin for "no U," usual error bars are given for "no flip." For both temperatures, the LB cur-

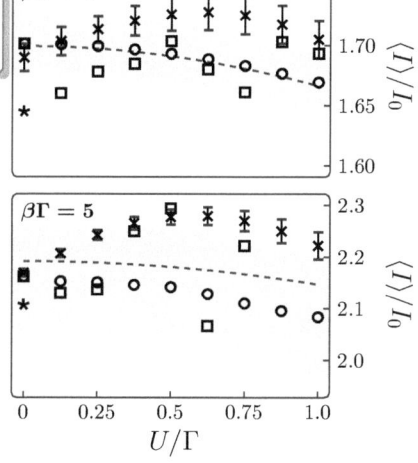

rent and the current without Coulomb scattering show only a weak dependence on U, as they only "feel" the single-particle energy shift. The current with full Coulomb interaction but fixed impurity ($\Longrightarrow$ effective magnetic field) shows a local maximum for $U \sim \Gamma/2$, which is more pronounced for $\beta\Gamma = 5$. All current values lie close together for small Coulomb energy, indicating again that flip-flip processes do not strongly affect the dot's resistivity. In particular, the LB and "no U" curves in **(a)** are in very good agreement. For the lower temperature, however, small differences between the current with flip-flop scattering and the LB values can be seen and seem to increase with the Coulomb interaction (size of energy shift $U/2$ in case of "no U"). Whether this feature is a genuine physical effect or an numerical artefact, remains inconclusive, as the LB values lie within the error margin of "no U" (about 5%). The ISPI values for the fully interacting dot (squares) vary strongly over the considered U interval, but are scattered around the "no flips" and "no U" curves. We refrain from giving error bars for these values (see discussion, also concerning the red crosses at $U = 0$).

current from its value at $U = 0$ to its maximum is significantly higher for the fully interacting dot (squares and star) than for the dot without flip-flops (crosses), it seems that mixed coherent (higher-order) scattering processes of both the Coulomb- and the electron-impurity interaction are necessary to fully understand and predict the nonequilibrium dynamics in this deep quantum regime.

5.5. Summary

This chapter contained a first systematic investigation of the long-time dynamics of an initially polarised impurity spin in a SLQD that is coupled to unpolarised, metallic leads, in the deep quantum regime, while taking into account all relevant correlations. We were able to describe the simulated exponential decay of the impurity polarisation by a simple rate equation and used the real-time diagrammatic technique by Schoeller and Schön [46] to derive an analytical expression for relaxation rates that implement the effects of sequential flip-flop processes in the system. For most of the studied parameter configurations, sequential flip-flops turn out to be the prevalent cause of the impurity spin relaxation, which can be deduced from the nearly quadratic dependence of the simulated, fully coherent inverse relaxation times on the interaction strength J. This allowed us to develop an intuitive explanation of the observed relaxation dependence on various system parameters, based on two-electron tunnelling processes (figure 5.9).

Yet, although sequential theory and numerical results turn out to agree quite well in some situations, the sequential theory never exceeds the status of a crude, first approximation of the dynamics in the inherently nonperturbative regime studied here. As soon as both interaction strengths U and J are of order $\sim \Gamma$ (especially so for temperatures as low as $\beta\Gamma = 5$), the predicted behaviour of the fully coherent ISPI values and the sequential approximation can be diametrically opposed, while the discrepancies between them can be as large as 100%. This can be attributed to the various

5. Spin-Relaxation by a Charge Current: Exact Results

assumptions needed to derive the sequential flip-flop rates.

Along the way, we analysed the (stationary) charge current and how it is affected by the electron-impurity interaction. Since the leads are unpolarised and the current is not sensible to the spin orientation of a tunnelling electron, the flip-flop component $\hat{H}_{\text{int}}^{\perp}$ does not largely affect the current. Within the margin of error, the decrease of $\langle I \rangle$ with growing J is solely caused by an effective shift of the single-particle dot energies due to

the longitudinal part of the interaction. This was checked (see figure 5.4) both by "turning-off" the flip-flops in the simulation (by setting $m_{\max}$ to zero) and a comparison to an approximate Landauer-Büttiker current based on the free Green's function $(G_{0,\sigma}^{\text{el},J})_\omega$ from equation (5.7). The LB current is systematically higher than the ISPI values but yields, nevertheless, often viable first approximations to the ISPI results. Our study of intermediate to strong interaction regimes (see figure 5.12), however, revealed several interesting signatures of correlations effects, which cannot be reproduced by our mean-field-type LB expression. It seems that at lower temperatures and in combination with the Coulomb interaction, flip-flop processes can have a (small) effect on the dot's resistivity. More importantly, we saw that Coulomb correlations can increase the current compared to the incoherent case and might have observed an interplay of both Coulomb- and impurity correlations both for $\beta\Gamma = 1$ and $\beta\Gamma = 5$. Further optimisations provided, the ISPI scheme might help clarifying these fundamental and fascinating questions in the future.

6

Conclusions and Outlook

GENERAL TOPIC OF this work is the nonequilibrium transport of particles, charge, and spin through an idealised, fundamental example of an open quantum system: a single-level quantum dot (SLQD) in contact with macroscopic leads. The thesis is divided into three main parts: The study of certain transport features in the Coulomb-blockade regime in chapter 3 is followed by a description of how to adopt the numerical exact, deterministic method of the *iterative summation of path integrals* (ISPI) by Weiss et al. [120] to the SLQD with an additional magnetic impurity in chapter 4. After that, this fully coherent numerical scheme is used to investigate the real-time impurity spin- and current dynamics in the deep quantum regime, in which all the system energies are of the same order of magnitude.

In chapter 3, the diagrammatic technique by Schoeller and Schön [46] provides us with Markovian rate equations, which are used to calculate the stationary density matrix of the SLQD (without impurity) and charge current flowing between the leads. As we are interested in the Coulomb-blockade regime, the tunnel-coupling Γ has to be weak and the temperature small (more precisely, $\Gamma \ll \beta^{-1} \ll U$, where U is the Coulomb energy). The rate equations are then expanded in orders of Γ, where the first- and second-order terms that are taken into account describe sequential and cotunnelling transport, respectively. In the Coulomb-blockade valley

6. Conclusions and Outlook

associated to single-particle occupation of the dot, the sequential current is exponentially suppressed and second-order cotunnelling dominates the transport.[1] It corresponds to the set of voltage configurations, for which both lead's electrochemical potentials lie between the (energetically) lowest and highest so-called transport channels (energy distances between dot states that differ by one in particle number). In case the two single-electron energies differ by Δ with $0 < \Delta < U/2$, this Coulomb-blockade (or cotunnelling-) valley exhibits a rich internal structure, which is shown in figure 3.1 and gives rise to a complex transport behaviour.

For $|eV_{\text{bias}}| < \Delta$, only elastic cotunnelling is present, the current is small, and the dot is in the spin state with the lower energy (the single-particle ground state). If $|eV_{\text{bias}}| > \Delta$, however, and as long as both chemical potentials μ_p (with $p = \text{L}, \text{R}$) lie (i) above the empty-dot channels and (ii) below the channels for tunnelling in and out of the doubly occupied state, inelastic cotunnelling provides for a finite occupation of the spin state with higher energy (excited state). Finally, for an increasing bias this occupation is reduced due to sequential transport, once at least one of the channels for tunnelling with the excited-state is in resonance with or lies between the μ_p. Among other things, this cotunnelling-mediated sequential transport manifests in a current step close to the resonance of an excited-state channel with a lead's chemical potential or, equivalently, in a peak of the differential conductance dI/dV_{bias}.

We focus on this conductance peak by studying its dependence on a phenomenological rate θ for spin relaxation between the excited- and ground state. When tuning θ between zero and infinity, we first find the conductance peak growing, from its finite size for vanishing relaxation, to assume a maximum height for $\theta \simeq \Gamma/2$. Faster relaxation lets it shrink again and

---

[1]This is, in fact, true for all Coulomb-blockade valleys. For the SLQD, there exist three valleys, one for each occupation number. All valleys together form the Coulomb-blockade regime.

6. Conclusions and Outlook

completely vanish in the limit $\theta \to \infty$. This can be explained by the fact that the current on both "sides" of the conductance peak (corresponding to bias voltages smaller or larger than the peak's position) is affected differently by the spin relaxation. On the smaller-bias side, the current and the dot's occupation are solely determined by cotunnelling contributions, which scale roughly as Γ^2/U. For larger bias, however, the spin relaxation competes with sequential tunnelling proportional to Γ. In both cases, the current decreases with growing θ. Hence, while for $\theta \simeq \Gamma/2$, the current is already almost maximally diminished on the small-bias side of the peak, it is still hardly affected by the relaxation for larger voltages. As a result, the conductance peak reaches its maximum height for this small but considerable relaxation. This phenomenon could provide means to directly influence the single-particle occupations in experiments and allows to facilitate measurements of excited state resonances by adjusting either the tunnel coupling Γ or the relaxation rate θ.

In chapter 4, the ISPI method is adopted to a SLQD that contains a quantum mechanical spin-1/2 impurity. To do so, we shortly review the notions of Grassmann numbers and fermionic coherent states, which are subsequently used to derive a (time-discrete) path integral representation of the Keldysh partition function $\mathcal{Z}$. In the course of the derivation, a discrete Hubbard-Stratonovich transformation maps the model with on-dot Coulomb repulsion to a system of free particles that interact with fluctuating, Ising-like virtual spin fields. By adding suitable source terms to the system's action, we obtain a generating function from $\mathcal{Z}$ that allows to calculate expectation values of arbitrary observables. Integrating over all electronic degrees of freedom eventually yields an expression of this generating function that is basically a discrete path sum over the determinant of the (path-dependent) inverse Green's matrix.

6. Conclusions and Outlook

To perform this sum, whose number of summands grows exponentially with propagation time, the actual ISPI scheme is applied. It allows to immensely reduce the computational costs by systematically neglecting irrelevant (long-) time correlations. The procedure is based on (i) the observation that correlations in the macroscopic leads decay exponentially with some finite memory time and (ii) the assumption that the on-dot correlations are limited to essentially the same value by virtue of the tunnel coupling. For the sake of numerical stability, we show in section 4.2.3 how to deal with (cancel) divergent elements of the discrete inverse Green's matrix without affecting the validity of the expressions. The resulting generating function is afflicted with systematical errors; in addition to the step-width δ_t of the time-discretization also the system's memory time τ_c is finite. Nevertheless, exact (numerically) values can be obtained with appropriate extrapolation schemes that numerically implement the limits $\delta_t \to 0$ and $\tau_c \to \infty$. Such schemes are reviewed in section 4.2.4 along with a restriction of the path sum to impurity paths with less than maximum number of flips. While not being a part of the original ISPI scheme, the latter is, if convergent, capable of further reducing the numerical costs of the simulation by at least one order of magnitude.

Finally, chapter 5 contains a systematical analysis both of the stationary current and the dynamics of the impurity spin orientation as functions of the interaction strengths J and U, the bias voltage V_{bias} and, in less detail, the temperature T. It is focused on the essentially non-perturbative, deep quantum regime that is characterised by the absence of a small model parameter. For all considered sets of parameters, we observe an exponential decay of the impurity spin, during propagation times of tens or even hundreds of $\hbar/\Gamma$, if its initial state is polarised. In contrast to this slow relaxation, the current assumes its stationary value at the latest after $\sim \hbar/\Gamma$ and can, for our purposes, be considered as time-independent. The simplest

6. Conclusions and Outlook

ansatz that reproduces the exponential decay of the impurity spin is a rate equation, where the rates for impurity flip processes are constant in time. To get a (more intuitive) understanding of the physics behind these simulated results, we compare them to analytical expressions (of the current and the relaxation rate), which are both fairly accurate and easy to interpret. Note, that there are some similarities between the systems and situations that are described in chapters 3 and 5. In both cases, the current through the dot is influenced by some on-dot relaxation processes and the ensuing interplay between correlated dynamics and decoherence leads to interesting signatures in the mesoscopic system. In chapter 3, the correlated *tunnelling current* is influenced by a relaxation of the electron spin due to the coupling to some unspecified, *dissipative* bath. In chapter 5, it is the *coherent* spin dynamics due to the electron-impurity interaction that couples the impurity to the tunnelling electrons from the *macroscopic leads*, which leads to the impurity spin relaxation. Hence, both chapters have in common, that they discuss some interplay of correlated quantum dynamics with decoherence effects. On the other hand, as the considered transport regimes and the respective sources of relaxation differ in both cases, the observed effects are not directly related or comparable.

In case of the ISPI current, we consider a slightly modified variant of the result from Landauer and Büttiker [202], which includes the Coulomb- and electron-impurity interaction solely via shifts of the single-particle energies. The observed reasonable agreement between both current values allows us to explain essential qualitative features of the simulated current with the help of an intuitive picture in terms of tunnelling electrons. We conclude that, as a first approximation, the charge current is only affected by the longitudinal part of the electron-impurity interaction. The numerical rate for

6. Conclusions and Outlook

impurity spin flips[2] we compare to the analytical rate of *sequential flips*, which is constructed with the help of Schoeller's and Schön's real-time diagrammatic technique. In most of the parameter configurations studied, this lowest-order approximation in J is of the correct order of magnitude and is influenced similarly by variations of the model parameters. For vanishing U and $J \lesssim \Gamma/2$, the resulting relaxation times versus V_{bias} even show a compelling qualitative agreement. Hence, we assume that in this case a reasonable understanding of the physics at hand is conveyed by the intuitive picture of sequential impurity spin flips. These complex processes are composed of two (sequential) single-electron tunnellings and a flip-flop process; the corresponding rates are given by the energy integral over the probability density for the two (sequential) tunnelling processes to occur.

Although the LB current and the sequential flip rates deliver some first insight into the processes determining current and relaxation, they do not, by any means, exhaustively explain the simulation results. The sequential relaxation rate is essentially a perturbative entity and, as in the considered transport regime interaction strength J is not a small parameter of the system, higher-order flip-flop processes contribute considerably to the relaxation. Furthermore, accuracy and adequacy of both the LB current and the sequential rate suffer from the fact that they are based on a phenomenological, free Green's function that includes the interactions via single-particle energy shifts. In contrast to that, the ISPI simulation fully accounts for all the on-dot interactions, relevant time correlations, and real-time dependencies. The systematic exploration of the parameter space reveals that both the quantitative and qualitative deviations between the analytical approximations and the numerical simulation increase with growing on-dot interactions and decreasing temperatures.

---

[2]The underlying elementary processes are flip-flops between an electron- and an impurity spin.

6. Conclusions and Outlook

With this last chapter, we merely get a first glimpse of the interesting, new physics that comes into reach with the ISPI scheme. Already for the presented investigation of the impurity spin relaxation, a number of questions are left out of consideration, so far. How precisely does the inverse relaxation time depend on the temperature? How does an external magnetic field or a shift in the gate voltage affect the decay time? Is it possible to have a finite stationary polarisation in case of a ferromagnetic electron-impurity interaction ($J < 0$)? Can we (clearly) detect a time-dependence of the relaxation rates themselves and under which conditions? These are just a few of the many questions, that can be asked in the general context of chapter 5. But we do not have to stop there. For example, the simulation can easily be extended to calculate the x- and y-components of the impurity spin, thus yielding the complete coherent spin dynamics on the Bloch sphere. This would be particularly interesting, if an oscillating term $\propto \hat{\tau}_{x,y}$ were added to the impurity Hamiltonian as in the work of Mitra and Millis [204, 205]. This would pave the way for studying the real-time dephasing of the oscillating impurity spin and the violation of the fluctuation dissipation theorem in the presence of the longitudinal *and* the flip-flop interaction. Also, it might be worthwhile to consider the time-dependent spin current that is caused by the decay of the impurity spin.

As a more technical point, the simulation's efficiency and accuracy would benefit considerably from a way to cancel the divergent terms in the original, real-time version of the generating function that does not require multiplying it with the free Green's function. Particularly the latter is still an issue with current computers and for intermediate interactions. Not only would such a scheme expand the frontiers of the treatable parameter range towards stronger interactions (mostly due to reduced running times). It also allowed to implement an arbitrary real-time dependence of model parameters like the tunnel coupling or the bias- and gate voltages. Possible follow-up mat-

6. Conclusions and Outlook

ters to investigate include the application of pulsed or ac voltages and tunnel coupling. In the medium term, it would be interesting to look at a model containing two or more impurities, as electrons in a small quantum dot can mediate a ferromagnetic interaction and, as a consequence, a finite magnetisation between embedded manganese impurities [99–103]. Accordingly, we would expect to see similar effects for a SLQD with two or more spin-1/2 impurities, while the ISPI simulation allowed to scrutinise the real-time dynamics and all-electrical control of the dot's magnetisation. Due to the numerical costs increasing with system size, however, in view of today's computational capacities, adding another impurity to the dot would very likely require to set the Coulomb interaction either to zero or to infinity.

A

Evolution Operator and Von Neumann Equation

It is the defining property of the time evolution operator $\hat{U}(t, t')$ for a system with Hamiltonian $\hat{H}(t)$, that it connects Schrödinger states at times t and t' by

$$|\psi(t)\rangle = \hat{U}(t, t')|\psi(t')\rangle. \tag{A.1}$$

Since the normalization of the states has to be preserved, $\hat{U}$ has to be unitary and from the case $t = t'$ follows $\hat{U}(t, t) \equiv \hat{\mathbb{1}}$. Furthermore, we have

$$\hat{U}(t, t')\hat{U}(t', t) = \hat{U}(t, t) = \hat{\mathbb{1}} \implies \hat{U}(t', t) = \hat{U}^{-1}(t, t') \equiv \hat{U}^\dagger(t, t'). \tag{A.2}$$

By plugging (A.1) into the Schrödinger equation $i\hbar\partial_t|\psi(t)\rangle = \hat{H}(t)|\psi(t)\rangle$, we arrive at the following equation of motion

$$i\hbar\partial_t \hat{U}(t, t') = \hat{H}(t)\hat{U}(t, t'), \tag{A.3}$$

which can be solved formally by separation of variables and iterated integration:

$$\hat{U}(t, t') = \hat{\mathbb{1}} - \frac{i}{\hbar}\int_{t'}^{t} dt_1\, \hat{H}(t_1)\hat{U}(t_1, t') = \ldots$$

A. Evolution Operator and Von Neumann Equation

$$\begin{aligned}
&= \hat{\mathbb{1}} - \frac{i}{\hbar} \int_{t'}^{t} dt_1 \, \hat{H}(t_1) \left[\hat{\mathbb{1}} - \frac{i}{\hbar} \int_{t'}^{t_1} dt_2 \, \hat{H}(t_2) \hat{U}(t_2, t') \right] \\
&= \hat{\mathbb{1}} - \frac{i}{\hbar} \int_{t'}^{t} dt_1 \, \hat{H}(t_1) + \left(-\frac{i}{\hbar}\right)^2 \int_{t'}^{t} dt_1 \int_{t'}^{t_1} dt_2 \, \hat{H}(t_1) \hat{H}(t_2) \hat{U}(t_2, t') \\
&= \hat{\mathbb{1}} - \frac{i}{\hbar} \int_{t'}^{t} dt_1 \, \hat{H}(t_1) + \left(-\frac{i}{\hbar}\right)^2 \int_{t'}^{t} dt_1 \int_{t'}^{t_1} dt_2 \, \hat{H}(t_1) \hat{H}(t_2) + \ldots \\
&\quad + \left(-\frac{i}{\hbar}\right)^n \int_{t'}^{t} dt_1 \int_{t'}^{t_1} dt_2 \cdots \int_{t'}^{t_{n-1}} dt_n \, \hat{H}(t_1) \hat{H}(t_2) \cdots \hat{H}(t_n) + \ldots \\
&=: \hat{T} \exp\left(-\frac{i}{\hbar} \int_{t'}^{t} \hat{H}(\tilde{t}) d\tilde{t}\right).
\end{aligned}$$

(A.4)

With the last step, a notation for the infinite series of integrals is defined. $\hat{T}$ is the time ordering operator as introduced in section 2.2.2.

That $\hat{\rho}(t) = \hat{U}(t, t') \hat{\rho}(t') \hat{U}^\dagger(t', t)$ is indeed the formal solution of the Von Neumann equation $\partial_t \hat{\rho}(t) = -i/\hbar [\hat{H}(t), \hat{\rho}(t)]$, can be checked by just inserting the former into the latter and using the above mentioned properties of $\hat{U}$.

$$\begin{aligned}
i\hbar \partial_t \hat{\rho}(t) &= i\hbar \partial_t [\hat{U}(t, t') \hat{\rho}(t') \hat{U}^\dagger(t', t)] \\
&= [i\hbar \partial_t \hat{U}(t, t')] \hat{\rho}(t') \hat{U}^\dagger(t', t) + \hat{U}(t, t') \hat{\rho}(t') [i\hbar \partial_t \hat{U}^\dagger(t', t)] \\
&= \hat{H}(t) \hat{U}(t, t') \hat{\rho}(t') \hat{U}^\dagger(t', t) - \hat{U}(t, t') \hat{\rho}(t') [\hat{H}(t) \hat{U}(t, t')]^\dagger \quad \text{(A.5)} \\
&= [\hat{H}(t), \hat{U}(t, t') \hat{\rho}(t') \hat{U}^\dagger(t', t)] \\
&= [\hat{H}(t), \hat{\rho}(t)].
\end{aligned}$$

In the step from the second to third row, we used the Schrödinger equation (A.3).

B

Sequential Rates: Tracing out the Leads

We start from equation (2.21), insert the tunneling Hamiltonian $\hat{H}_T$ from (2.1f) and write the states $|\psi\rangle$ and $|\psi'\rangle$ explicitly as tensor product states. With the spin σ that is uniquely determined by the choice of χ and χ', this yields

$$\frac{\hbar}{2\pi|\gamma|^2} W_{\chi \leftarrow \chi'}$$
$$= \sum_{\substack{\boldsymbol{k}_L \boldsymbol{k}_R \; \boldsymbol{k}_1 p_1 \\ \boldsymbol{k}'_L \boldsymbol{k}'_R \; \boldsymbol{k}_2 p_2}} \langle \chi' \boldsymbol{k}'_L \boldsymbol{k}'_R | \hat{d}^\dagger_\sigma \hat{c}_{\boldsymbol{k}_1 p_1} + \hat{c}^\dagger_{\boldsymbol{k}_1 p_1} \hat{d}_\sigma | \chi \boldsymbol{k}_L \boldsymbol{k}_R \rangle \times \quad\quad \text{(B.1)}$$
$$\langle \chi \boldsymbol{k}_L \boldsymbol{k}_R | (\hat{d}^\dagger_\sigma \hat{c}_{\boldsymbol{k}_2 p_2} + \hat{c}^\dagger_{\boldsymbol{k}_2 p_2} \hat{d}_\sigma) \hat{\rho}_{\text{leads}} | \chi' \boldsymbol{k}'_L \boldsymbol{k}'_R \rangle \delta(E_\psi - E_{\psi'}).$$

Off all four termes, that would result from expanding the operator sums, only those two contribute that eventually conserve the particle number. This is because the operator products are sandwiched between the pair of dual states $\langle \psi'|$ and $|\psi'\rangle$. The corresponding summation is equivalent to a trace over the leads. We arrive at

$$W_{\chi \leftarrow \chi'} \propto$$
$$\sum_{\boldsymbol{k}_L \boldsymbol{k}_R} \sum_{\substack{\boldsymbol{k}_1 p_1 \\ \boldsymbol{k}_2 p_2}} \Big(\; |\langle \chi | \hat{d}^\dagger_\sigma | \chi' \rangle|^2 \; \text{Tr}_{\text{leads}} \{ \hat{c}^\dagger_{\boldsymbol{k}_1 p_1} |\boldsymbol{k}_L \boldsymbol{k}_R\rangle \langle \boldsymbol{k}_L \boldsymbol{k}_R | \hat{c}_{\boldsymbol{k}_2 p_2} \hat{\rho}_{\text{leads}} \}$$
$$+ |\langle \chi | \hat{d}_\sigma | \chi' \rangle|^2 \; \text{Tr}_{\text{leads}} \{ \hat{c}_{\boldsymbol{k}_1 p_1} |\boldsymbol{k}_L \boldsymbol{k}_R\rangle \langle \boldsymbol{k}_L \boldsymbol{k}_R | \hat{c}^\dagger_{\boldsymbol{k}_2 p_2} \hat{\rho}_{\text{leads}} \} \Big) \times \quad \text{(B.2)}$$
$$\times \delta(E_\psi - E_{\psi'}).$$

B. Sequential Rates: Tracing out the Leads

The remaining two terms for tunneling out off and onto the dot, respectively, can be further evaluated, if we consider the restrictions for the values of the variables that can be read of the operator products in each of the traces. Clearly, the leads involved in each tunneling process have to be equal: $p_1 = p_2 = p$. Hence, off the outer sums, only the one over $\boldsymbol{k}_p$ has to be considered. Furthermore, depending on p we can derive the conditions $\epsilon_{\boldsymbol{k}_p} - \epsilon_{\boldsymbol{k}'_p} = \mp\epsilon_{\boldsymbol{k}_2}$ and $\epsilon_{\boldsymbol{k}_{-p}} - \epsilon_{\boldsymbol{k}'_{-p}} = 0$. Finally, it can be concluded, that $\boldsymbol{k}_1 = \boldsymbol{k}_2$. By taking all those constraints into account, we arrive at

$$W_{\chi \leftarrow \chi'} \propto \ |\langle\chi|\hat{d}^\dagger_\sigma|\chi'\rangle|^2 \sum_{kp} \mathrm{Tr}_p\{\hat{c}^\dagger_{kp}\hat{c}_{kp}\hat{\rho}_p\}\delta(\epsilon_\chi - \epsilon_{\chi'} - \epsilon_k) \\ + |\langle\chi|\hat{d}_\sigma|\chi'\rangle|^2 \sum_{kp} \mathrm{Tr}_p\{\hat{c}_{kp}\hat{c}^\dagger_{kp}\hat{\rho}_p\}\delta(\epsilon_\chi - \epsilon_{\chi'} + \epsilon_k).$$ (B.3)

Now, we can identify $\hat{c}^\dagger_{kp}\hat{c}_{kp} = \hat{n}_{kp}$ and $\hat{c}_{kp}\hat{c}^\dagger_{kp} = \hat{1} - \hat{n}_{kp}$ and make use of the fact, that the expectation value of the particle number operator $\langle \hat{n}_{kp} \rangle :=$ $\mathrm{Tr}_p\{\hat{\rho}_p \hat{n}_{kp}\}$ with wave vector $\boldsymbol{k}$ of lead p is given by the Fermi function $f(\epsilon_k - \mu_p) = [1+\exp(\beta\epsilon_{kp})]^{-1}$, the thermal occupation function for a free gas of fermions. Here, we used $\epsilon_{kp} = \epsilon_k - \mu_p$. To carry out the remaining sum over $\boldsymbol{k}$, we transform it into an integration over energy space (see section 2.3.1) in the wide-band limit: $\sum_k \to \varrho(\epsilon_F) \int d\epsilon_k$, where $\varrho(\epsilon_F)$ is the density of states at the Fermi level ϵ_F. In the end, this yields expression (2.23).

C

The Interaction Picture

The interaction picture is useful to conceptually separate a Hamiltonian $\hat{H}$ into parts $\hat{H}_0$ and $\hat{H}_1$, which are often called the free and the interaction part, respectively. It is a mixture of the Schrödinger picture, in which the time dependence is carried by the Hilbert states, and the Heisenberg picture, where it is assigned to the operators. In the interaction picture, operators propagate in time according to the free evolution part only. The Hilbert states carry the "remaining" time dependence, which is given by the condition that the resulting expectation values are equivalent to the Schrödinger picture. Since all relations between states and operators are essentially differential equations in time, a reference point (initial condition) t_i has to be fixed. In this point, the Schrödinger and interaction picture coincide. In the following, we denote operators and Hilbert states in the different pictures with the superscripts "S" and "I" for the Schrödinger and interaction picture, respectively. Thus, we obtain the defining relations

$$|\psi^S(t)\rangle = \hat{U}(t,t_i)|\psi^S(t_i)\rangle, \quad |\psi^I(t)\rangle = \hat{U}^I(t,t_i)|\psi^I(t_i)\rangle$$
$$|\psi^S(t_i)\rangle = |\psi^I(t_i)\rangle,$$
$$\hat{O}^S = \hat{O}^I(t_i) \quad \text{and} \quad \hat{O}^I(t) = \hat{U}_0(t_i,t)\hat{O}^S\hat{U}_0(t,t_i) \quad \text{(C.1)}$$
$$\langle\psi^I(t)|\hat{O}^I(t)|\psi^I(t)\rangle = \langle\psi^S(t)|\hat{O}^S|\psi^S(t)\rangle,$$

where $\hat{U}$ is the Schrödinger full evolution operator (A.4) and $\hat{U}_0$ the free evolution operator, which is the result of replacing $\hat{H}$ with $\hat{H}_0$ in $\hat{U}$. The

C. The Interaction Picture

interaction picture propagator $\hat{U}^I$ can then be determined by the equations (C.1) to be

$$\hat{U}_I(t, t_i) = \hat{U}_0(t_i, t)\hat{U}(t, t_i). \tag{C.2}$$

To derive an integral expression of $\hat{U}^I$ similar to (A.4), we form a Schrödinger equation by deriving (C.2) with respect to time and use (A.3):

$$\begin{aligned}
i\hbar\partial_t \hat{U}^I(t, t_i) &= i\hbar\partial_t[\hat{U}_0(t_i, t)\hat{U}(t, t_i)] \\
&= i\hbar[\{\partial_t\hat{U}_0(t_i, t)\}\hat{U}(t, t_i) + \hat{U}_0(t_i, t)\partial_t\hat{U}(t, t_i)] \\
&= [-\hat{H}_0\hat{U}_0(t_i, t)\hat{U}(t, t_i) + \hat{U}_0(t_i, t)\hat{H}\hat{U}(t, t_i)] \\
&= \hat{U}_0(t_i, t)[\hat{H} - \hat{H}_0]\hat{U}_0(t, t_i)\hat{U}_0(t_i, t)\hat{U}(t, t_i) \\
&= \hat{H}_1^I(t)\hat{U}^I(t, t_i).
\end{aligned} \tag{C.3}$$

Completely analogous to appendix A, from this equation we obtain the identity

$$\hat{U}^I(t, t') = \hat{T}\exp\left(-\frac{i}{\hbar}\int_{t'}^{t}\hat{H}_1^I(\tilde{t})\,d\tilde{t}\right). \tag{C.4}$$

We can use this to derive the alternative version of the full evolution operator, that was used in equation (4.6). With (C.2), we can write

$$\begin{aligned}
\hat{U}(t, t') &= \hat{U}_0(t, t_i)\hat{U}^I(t, t')\hat{U}_0(t_i, t') \\
&= \hat{U}_0(t, t_i)\hat{T}\exp\left(-\frac{i}{\hbar}\int_{t'}^{t}\hat{H}_1^I(\tilde{t})\,d\tilde{t}\right)\hat{U}_0(t_i, t') \\
&= \hat{U}_0(t, t') - \hat{U}_0(t, t_i)\frac{i}{\hbar}\int_{t'}^{t}dt_2\,\hat{H}_1^I(t_2)\,\hat{U}_0(t_i, t') \\
&\quad + \hat{U}_0(t, t_i)\left(-\frac{i}{\hbar}\right)^2\int_{t'}^{t}dt_2\int_{t_2}^{t}dt_3\,\hat{H}_1^I(t_3)\hat{H}_1^I(t_2)\,\hat{U}_0(t_i, t') + \ldots
\end{aligned}$$

C. The Interaction Picture

$$
\begin{aligned}
= \hat{U}_0(t, t') &+ \int_{t'}^{t} dt_2\, \hat{U}_0(t, t_2)\left(-\frac{i}{\hbar}\hat{H}_1\right)\hat{U}_0(t_2, t') \\
&+ \int_{t'}^{t} dt_2 \int_{t_2}^{t} dt_3\, \hat{U}_0(t, t_3)\left(-\frac{i}{\hbar}\hat{H}_1\right)\hat{U}_0(t_3, t_2) \times \\
&\quad \cdots \times \left(-\frac{i}{\hbar}\hat{H}_1\right)\hat{U}_0(t_2, t') + \ldots \quad (C.5) \\
= \sum_{N=2}^{\infty} \int_{t'}^{t} \cdots & \int_{t_{N-2}}^{t} \hat{U}_0(\Delta_t^{(N-1)})\left(-\frac{i}{\hbar}\hat{H}_1 dt_{N-1}\right)\cdots \hat{U}_0(\Delta_t^{(2)}) \times \\
&\quad \cdots \times \left(-\frac{i}{\hbar}\hat{H}_1 dt_2\right)\hat{U}_0(\Delta_t^{(1)})
\end{aligned}
$$

with $\Delta_t^{(k)} := t_{k+1} - t_k$ with $t_1 := t'$ and $t_N := t$.

D

Coherent States, Grassmann Numbers and Gaussian Integration

Partition of Unity

We check that (4.21) is indeed a partition of unity by plugging-in the corresponding expressions for the coherent states (4.16) and the exponential function (4.17):

$$\iint d\bar{f}\, df\, e^{-\bar{f}f} |f\rangle\langle f|$$
$$= \iint d\bar{f}\, df\, (1 - \bar{f}f)(\hat{\mathbb{1}} - f\hat{d}^\dagger)|0\rangle\langle 0|(\hat{\mathbb{1}} - \tilde{d}\bar{f})$$
$$= \iint d\bar{f}\, df\, \{|0\rangle\langle 0| - f|1\rangle\langle 0| - \bar{f}|0\rangle\langle 1| - \bar{f}f|0\rangle\langle 0| + f\bar{f}|1\rangle\langle 1|$$
$$\quad + \underbrace{\bar{f}ff}_{0}|1\rangle\langle 0| + \underbrace{f\bar{f}\bar{f}}_{0}|0\rangle\langle 1| - \underbrace{\bar{f}f f\bar{f}}_{0}|1\rangle\langle 1|\} \qquad (D.1)$$
$$= \iint d\bar{f}\, df\, \{|0\rangle\langle 0| - f|1\rangle\langle 0| - \bar{f}|0\rangle\langle 1| + f\bar{f}(|0\rangle\langle 0| + |1\rangle\langle 1|)\}$$
$$= \underbrace{(|0\rangle\langle 0| + |1\rangle\langle 1|)}_{\hat{\mathbb{1}}} \iint d\bar{f}\, df\, f\bar{f} = \hat{\mathbb{1}} \int d\bar{f}\, \bar{f} = \hat{\mathbb{1}}.$$

D. Coherent States, Grassmann Numbers and Gaussian Integration

In the second step (second to third line), we exploited the fact that in a Grassmann algebra with n generators, polynomial terms larger than n vanish, for they contain at least two equal factors. Either way, the Grassman integration over both generators would have eliminated them anyways, since all integrands that do not contain both generators (exactly once) vanish. Hence, all terms with less than two Grassmann numbers evaluate to zero, too, and we are left with the unity operator $\hat{\mathbb{1}}$. This prove can be extended to bases of multi-fermion coherent states and equation (4.22b). A straightforward version is shown by Negele and Orland [186].

Exponential Law

As an example for the fact, that the Grassmann analogies for usual real and complex exponential expressions hold, when Grassmann numbers appear pairwise, we prove equation (4.23):

$$\exp\left\{\sum_{i=1}^{n} \bar{f}_i f_i\right\}$$
$$= 1 + \sum_{i=1}^{n} \bar{f}_i f_i + \frac{1}{2!}\left[\sum_{i=1}^{n} \bar{f}_i f_i\right]^2 + \frac{1}{3!}\left[\sum_{i=1}^{n} \bar{f}_i f_i\right]^3 + \cdots \frac{1}{n!}\left[\sum_{i=1}^{n} \bar{f}_i f_i\right]^n$$
$$= 1 + \sum_{i=1}^{n} \bar{f}_i f_i + \frac{1}{2!}\sum_{j \neq k} \bar{f}_j f_j \bar{f}_k f_k + \frac{1}{3!} \sum_{j \neq k \neq l} \bar{f}_j f_j \bar{f}_k f_k \bar{f}_l f_l + \cdots$$
$$+ \frac{1}{n!} \sum_{i_1 \neq i_2 \neq \ldots \neq i_n} \bar{f}_{i_1} f_{i_1} \bar{f}_{i_2} f_{i_2} \cdots \bar{f}_{i_n} f_{i_n} \tag{D.2}$$
$$= 1 + \sum_{i=1}^{n} \bar{f}_i f_i + \frac{1}{2!}\sum_{j<k} \left[\bar{f}_j f_j \bar{f}_k f_k + \bar{f}_k f_k \bar{f}_j f_j\right]$$
$$+ \frac{1}{3!} \sum_{t \in T_n(3)} \sum_{p \in S_3} \left[\bar{f}_{tpt^{-1}(1)} f_{tpt^{-1}(1)} \bar{f}_{tpt^{-1}(2)} f_{tpt^{-1}(2)} \bar{f}_{tpt^{-1}(3)} f_{tpt^{-1}(3)}\right] + \cdots$$

D. Coherent States, Grassmann Numbers and Gaussian Integration

$$+ \ldots$$
$$+ \frac{1}{n!} \sum_{p \in S_n} \left[\bar{f}_{p(1)} f_{p(1)} \bar{f}_{p(2)} f_{p(2)} \cdots \bar{f}_{p(n)} f_{p(n)} \right]$$
$$= 1 + \sum_{i=1}^{n} \bar{f}_i f_i + \sum_{j<k} \bar{f}_j f_j \bar{f}_k f_k + \sum_{j<k<l} \bar{f}_j f_j \bar{f}_k f_k \bar{f}_l f_l + \ldots + \prod_{i=1}^{n} \bar{f}_i f_i$$
$$= (1 + \bar{f}_1 f_1)(1 + \bar{f}_2 f_2) \cdots (1 + \bar{f}_n f_n) = \prod_{i=1}^{n} \exp\{\bar{f}_i f_i\},$$

where $T_n(k)$ is the set of injective maps t of $\{1, 2, \ldots, k\}$ into $\{1, 2, \ldots, n\}$ with $t(1) < t(2) < \ldots < t(k)$ and S_k is the symmetric group of degree k. It is because Grassmann numbers with the same index appear in pairs, that no signs are collected by ordering all Grassmann products with ascending numbers of indices.

Derivation of Gaussian Path Integral

To derive the path integral expressions (4.26) explicitly, we insert the partition of unity (4.25) for the whole model (2.1) on both sides of every short time propagator in (4.9), assuming free propagation, $\hat{U}_{\delta_t} \equiv \,:\!\hat{U}_0(\delta_t)\!:\,$:

$$\hat{U}_0(t_f, t_i)$$
$$\approx \sum_{\tau_1 \ldots \tau_N} \int \prod_{j=1}^{N} [d\bar{\Psi}_j^\tau d\Psi_j^\tau] \, e^{-\sum_{j=1}^{N} \bar{\Psi}_i \Psi_i} \left[\prod_{j=1}^{N} \Psi_{j+1}^\tau | :\hat{U}_0(\delta_t): |\Psi_j^\tau \rangle \right] |\Psi_{N+1}^\tau\rangle\langle\Psi_1^\tau|$$
$$= \int \prod_{i=1}^{N} [d\bar{\Psi}_j d\Psi_j] \exp\left\{ i \sum_{i=2}^{N} \left(i\bar{\Psi}_j \frac{\Psi_j - \Psi_{j-1}}{\delta_t} - \frac{H_0}{\hbar}[\bar{\Psi}_j, \Psi_{j-1}, \tau_i] \right) \delta_t \right.$$
$$\left. \ldots - \bar{\Psi}_1 \Psi_1 \right\} |\Psi_{N+1}^\tau\rangle\langle\Psi_1^\tau|$$
$$\approx \int \prod_{i=1}^{N} [d\bar{\Psi}_j d\Psi_j] \exp\{i S_0[\{\bar{\Psi}, \Psi, \tau_i\}]\} |\Psi_{N+1}^\tau\rangle\langle\Psi_1^\tau|.$$

(D.3)

D. Coherent States, Grassmann Numbers and Gaussian Integration

The discrete action can be identified with

$$S_0[\{\bar{\Psi}, \Psi, \tau_i\}] = \sum_{i=2}^{N} \left(i\bar{\Psi}_j \frac{\Psi_j - \Psi_{j-1}}{\delta_t} - \frac{H_0^{\text{el}}}{\hbar}[\bar{\Psi}_j, \Psi_{j-1}, \tau_i] \right) \delta_t \qquad \text{(D.4)}$$

$$\underbrace{\cdots - \frac{\Delta_{\text{imp}} \Delta_t}{2\hbar} \tau_i}_{S^{\text{imp}}},$$

which, in the limit $\delta_t \to 0$, converges to the continuous expression (4.27). The approximative relation in the last line of (D.3) has to be used, since term $-\bar{\Psi}_1 \Psi_1$ is unaccounted for in the definition of the discrete action. In the limit of infinitely small δ_t, however, this boundary contribution that can be neglected. The term S^{imp} we derived here is equal to (4.29) for an impurity path $(\tau_i, \ldots, \tau_i)$ with constant orientation—the only one that can contribute in the case of free time evolution.

The Trace Operation

The proof of expression (4.43) shown here, is based on the explanations from [186]. Let $\{|k\rangle\}$ be the complete set of Fock states in a fermionic Hilbert space with n d.o.f. We use the coherent state partition of unity (4.22b) to rewrite the trace of an operator $\hat{O}$ in the Fock basis as

$$\begin{aligned}
\text{Tr}\{\hat{O}\} &= \sum_k \langle k|\hat{O}|k\rangle = \sum_k \int d\bar{\mathfrak{f}} \, d\mathfrak{f} \, e^{-\bar{\mathfrak{f}}\mathfrak{f}} \langle k|\mathfrak{f}\rangle \langle \mathfrak{f}|\hat{O}|k\rangle \\
&= \sum_{k,k'} \int d\bar{\mathfrak{f}} \, d\mathfrak{f} \, e^{-\bar{\mathfrak{f}}\mathfrak{f}} \langle 0|f_{i_1} \cdots f_{i_k}|\mathfrak{f}\rangle \langle \mathfrak{f}|\alpha(k')|k'\rangle \\
&= \sum_{k,k'} \int d\bar{\mathfrak{f}} \, d\mathfrak{f} \, e^{-\bar{\mathfrak{f}}\mathfrak{f}} \langle 0|f_{i_1} \cdots f_{i_k}|\mathfrak{f}\rangle \langle \mathfrak{f}|\alpha(k')\bar{f}_{j_1} \cdots \bar{f}_{j_{k'}}|0\rangle
\end{aligned}$$
(D.5)

D. Coherent States, Grassmann Numbers and Gaussian Integration

$$= \sum_{k,k'} \int d\bar{\mathfrak{F}} \, d\mathfrak{F} \, e^{-\bar{\mathfrak{F}}\mathfrak{F}} (-1)^{k'} \langle \mathfrak{F}|\alpha(k')\bar{f}_{j_1}\cdots\bar{f}_{j_{k'}}|0\rangle\langle 0|f_{i_1}\cdots f_{i_k}|\mathfrak{F}\rangle$$

$$= \sum_{k,k'} \int d\bar{\mathfrak{F}} \, d\mathfrak{F} \, e^{-\bar{\mathfrak{F}}\mathfrak{F}} \langle -\mathfrak{F}|\alpha(k')|k'\rangle\langle k|\mathfrak{F}\rangle$$

$$= \sum_{k} \int d\bar{\mathfrak{F}} \, d\mathfrak{F} \, e^{-\bar{\mathfrak{F}}\mathfrak{F}} \langle -\mathfrak{F}|\hat{O}|k\rangle\langle k|\mathfrak{F}\rangle$$

$$= \int d\bar{\mathfrak{F}} \, d\mathfrak{F} \, e^{-\bar{\mathfrak{F}}\mathfrak{F}} \langle -\mathfrak{F}|\hat{O}|\mathfrak{F}\rangle.$$

It is equally possible to attach the minus sign to the ket instead of the bra state. The crucial step is the one from line four to five. It is easy to see, that the correct sign for all possible numbers k' is automatically obtained, if a sign is attached to every Grassmann number in $\langle\mathfrak{F}|$.

Matrix Element of Thermal State

Equivalent to calculating (the fermionic part of) matrix element (4.49) for the *free, thermal* leads is to evaluate expression $\langle -\mathfrak{F}| \exp\{\sum_i \alpha_i \hat{n}_i\}|\mathfrak{F}\rangle$ with $\hat{n}_i = \hat{f}_i^\dagger \hat{f}_i$ of a general fermionic system with n d.o.f. As a first step, for a single d.o.f., we consider:

$$\langle f|e^{\alpha\hat{n}}|f\rangle$$
$$= \langle f|\hat{\mathbb{1}} + \sum_{k=1}^{\infty} \frac{\alpha^k}{k!} \hat{n}^k|f\rangle = \langle f|\hat{\mathbb{1}} + \hat{n}\sum_{k=1}^{\infty} \frac{\alpha^k}{k!}|f\rangle = \langle f|\hat{\mathbb{1}} + (e^\alpha - 1)\hat{n}|f\rangle \quad \text{(D.6)}$$
$$= [1 + (e^\alpha - 1)\bar{f}f]\langle f|f\rangle = [1 + (e^\alpha - 1)\bar{f}f](1 + \bar{f}f) = 1 + e^\alpha \bar{f}f$$
$$= \exp\{e^\alpha \bar{f}f\}.$$

Since for n differing fermions the particle number operators $\hat{n}_i$ commute pairwise, we can write $\exp\{\sum_i \alpha_i \hat{n}_i\} = \prod_i \exp\{\alpha_i \hat{n}_i\}$. With this, we get

D. Coherent States, Grassmann Numbers and Gaussian Integration

for the sought-after matrix element

$$
\begin{aligned}
&\langle -\mathfrak{F}| \exp\{\sum_i \alpha_i \hat{n}_i\}|\mathfrak{F}\rangle \\
&= \langle -\mathfrak{F}| \prod_i \exp\{\alpha_i \hat{n}_i\}|\mathfrak{F}\rangle = \langle -\mathfrak{F}| \prod_i [\hat{\mathbb{1}} + (e^{\alpha_i} - 1)\hat{n}_i]|\mathfrak{F}\rangle \\
&= \prod_i [1 - (e^{\alpha_i} - 1)\bar{f}_i f_i] \langle -\mathfrak{F}|\mathfrak{F}\rangle \\
&= \exp\{-\sum_i (e^{\alpha_i} - 1)\bar{f}_i f_i\} \exp\{-\sum_i \bar{f}_i f_i\} \\
&= \exp\{-\sum_i e^{\alpha_i} \bar{f}_i f_i\}.
\end{aligned}
\qquad (\text{D.7})
$$

In the last step, we used the exponential law as shown above. Replacing the summation index as $i \to k\sigma p$ and setting $\alpha_i \to -\beta(\epsilon_{k\sigma p} - \mu_p)$, immediately yields the result from equation (4.49).

E

Hubbard-Stratonovich Transformation

In the following, we review how the system of Coulomb interacting dot electrons ($\hat{H}^U_{\text{dot}}$) can be mapped to a system of non-interacting electrons, coupled to a field of Ising-like fluctuating spins, as described by equation (4.35). The dot's particle number operators obey the commutator relations $\hat{n}^2_\sigma = \hat{n}_\sigma$ and $[\hat{n}_\sigma, \hat{n}_{\sigma'}] = 0$. With these, we have

$$\forall_{n>0} : (\hat{n}_\uparrow - \hat{n}_\downarrow)^{2n} = (\hat{n}_\uparrow - \hat{n}_\downarrow)^2 \quad \text{and} \quad \forall_{n>0} : (\hat{n}_\uparrow - \hat{n}_\downarrow)^{2n-1} = (\hat{n}_\uparrow - \hat{n}_\downarrow).$$
(E.1)

Then, with $\Delta\hat{n} := \hat{n}_\uparrow - \hat{n}_\downarrow$, we get

$$\exp\{A\Delta\hat{n}^2\} = \sum_{m=0}^\infty \frac{A^m}{m!}\Delta\hat{n}^{2m} = \hat{1} + \Delta\hat{n}^2 \sum_{m=1}^\infty \frac{A^m}{m!} = \hat{1} + (e^A - 1)\Delta\hat{n}^2$$

(E.2a)

and

$$\exp\{B\Delta\hat{n}\} = \sum_{m=0}^{\infty} \frac{B^m}{m!}\Delta\hat{n}^m \tag{E.2b}$$

$$= \hat{1} + \left(\frac{B^1}{1!} + \frac{B^3}{3!} + \ldots\right)\Delta\hat{n} + \left(\frac{B^2}{2!} + \frac{B^4}{4!} + \ldots\right)\Delta\hat{n}^2$$

$$= \hat{1} + \sinh(B)\Delta\hat{n} + (\cosh(B) - 1)\Delta\hat{n}^2 \tag{E.2c}$$

Note that operator (E.2b) with the quadratic exponent has a term proportional to $\Delta\hat{n}$, whereas (E.2a) has not. To express the quartic exponential by means of the Gaussian one therefore requires, as a first step, to find a Gaussian operator lacking a $\Delta\hat{n}$-term. Since sinh is an odd function, while cosh is even, such an operator is given by the sum

$$\frac{1}{2}\left(e^{+B\Delta\hat{n}} - e^{-B\Delta\hat{n}}\right) = \frac{1}{2}\sum_{\zeta=\pm 1}\exp\{\zeta B\Delta\hat{n}\} = \hat{1} + (\cosh(B) - 1)\Delta\hat{n}^2. \tag{E.3}$$

It follows immediately, that the Gaussian (E.3) equals the quartic exponential operator (E.2a), if $\cosh(B) = e^A$. For a purely imaginary $A = \pm i|\Im A|$ and a complex number $B = |\Re B| \pm i|\Im B|$ with positive real part, we obtain the conditional equation

$$\cosh(|\Re B|)\cos(|\Im B|) \pm i \sinh(|\Re B|)\sin(|\Im B|)$$
$$= \cos(|\Im A|) \pm i \sin(|\Im A|), \tag{E.4}$$

where we used the addition theorem of the cosh function to arrive at the l.h.s.. For $0 \leq |\Im A| \leq \pi/2$ there exists the unique solution:

$$|\Re B| = \sinh^{-1}\sqrt{\sin(|\Im A|)} \quad \text{and} \quad |\Im B| = \sin^{-1}\sqrt{\sin(|\Im A|)}, \tag{E.5}$$

E. Hubbard-Stratonovich Transformation

which can be easily checked. We illustrate this for the real part of (E) by calculating the product of $\cosh(|\Re B|)$ and $\cos(|\Im B|)$:

$$\cosh(|\Re B|) = \sqrt{1 + \sinh^2(|\Re B|)} = \sqrt{1 + \sin(|\Im A|)}$$
$$\cos(|\Im B|) = \sqrt{1 - \sin^2(|\Im B|)} = \sqrt{1 - \sin(|\Im A|)} \qquad (\text{E.6})$$
$$\text{thus} \quad \cos(|\Im A|) = \sqrt{1 - \sin^2(|\Im A|)} = \cosh(|\Re B|)\cos(|\Im B|).$$

The imaginary part can be checked analogously. With this result, we can perform the desired mapping for our system. Using the identities $\hat{n}_\uparrow \hat{n}_\downarrow = \frac{1}{2}(\hat{n}_\uparrow + \hat{n}_\downarrow) - \frac{1}{2}(\hat{n}_\uparrow - \hat{n}_\downarrow)^2$, $\Im A = U\Delta_t/(2\hbar)$, and $B = \lambda \Delta_t$, we obtain equation (4.35) for the evolution operator $\exp\{-i/\hbar \hat{H}^U_{\text{dot}}\Delta_t\}$.

F

Lead Green's Function

Green's Function and Determinant

To evaluate the identities (4.59) and (4.60), we consider the discrete version of the free inverse Green's matrix G^{-1}_{leads} for the leads. It is worthwhile noticing that it is a block matrix, whose constituent blocks are all diagonal. Hence, by appropriate permutations of rows and columns, it can be sorted with respect the fermion d.o.f. In other words G^{-1}_{leads} is equivalent to the direct sum of inverse Green's functions g_R^{-1} of the individual lead electrons with $\boldsymbol{R} = (\boldsymbol{k}\sigma p)$:

$$G^{-1}_{\text{leads}} = \bigoplus_R g_R^{-1}. \tag{F.1}$$

It follows immediately, that the determinant and inverse of G^{-1}_{leads} are given by

$$\det\{G^{-1}_{\text{leads}}\} = \prod_R \det\{g_R^{-1}\} \quad \text{and} \quad G_{\text{leads}} = \bigoplus_R g_R. \tag{F.2}$$

Thus, all we are left with to arrive at (4.59) and (4.60), is to calculate determinant and inverse of

$$ig_R^{-1} = \begin{pmatrix} -1 & & & & & & & -\rho_0 \\ 1-i\phi & -1 & & & & & & \\ & 1-i\phi & -1 & & & & & \\ & & \ddots & \ddots & & & & \\ \hline & & & 1 & \ddots & & & \\ & & & & \ddots & -1 & & \\ & & & & & 1+i\phi & -1 & \\ & & & & & & 1+i\phi & -1 \end{pmatrix}, \tag{F.3}$$

241

F. Lead Green's Function

where $\rho_0 = \exp\{-\beta(\epsilon_k - \mu_p)\}$ and the lead's phase $\phi = \epsilon_k \delta_t/\hbar$ is independent of the impurity- and HS fields and, thus, of time. First, we calculate the determinant and obtain

$$\begin{aligned}
&\det\{ig_R^{-1}\}\\
&= \lim_{\delta_t \to 0} \det\{(-1)^{2N} + e^{-\beta(\epsilon_k - \mu_p)}(1 - i\epsilon_R \delta_t/\hbar)^{N-1}(1 + i\epsilon_R \delta_t/\hbar)^{N-1}\}\\
&= \lim_{\delta_t \to 0} \det\{(-1)^{2N} + e^{-\beta(\epsilon_k - \mu_p)}\{1 + (\epsilon_R \delta_t/\hbar)^2\}^{N-1}\} \quad (\text{F.4})\\
&\approx \lim_{\delta_t \to 0} \det\{1 + e^{-\beta(\epsilon_k - \mu_p)} \exp\{(\epsilon_R \delta_t/\hbar)^2 (N-1)\}\}\\
&= 1 + e^{-\beta(\epsilon_k - \mu_p)}.
\end{aligned}$$

Hence, $\det\{iG_{\text{leads}}^{-1}\} = \prod_R (1 + \exp\{-\beta(\epsilon_k - \mu_p)\})$. The additional sign in front of the i in (4.60) does not alter the value of the determinant, since the Green's matrix has an even dimension by construction. What is left to prove equation (4.60), is to show that this equals the constant $\mathcal{N}$. To this end, we define the many-particle lead states with m particles $|\psi_{\text{leads}}\rangle = |R_{i_1}, \ldots, R_{i_m}\rangle$, where the fermion states $|R_{i_k}\rangle$ with $1 \leq k \leq m$ are occupied and all other states are empty. Then we get:

$$\begin{aligned}
\mathcal{N} &= \text{Tr}[\exp\{-\beta \sum_p (\hat{H}_p - \mu_p \hat{N}_p)\}]\\
&= \sum_{\psi_{\text{leads}}} \langle \psi_{\text{leads}} | \exp\{-\beta \sum_p (\hat{H}_p - \mu_p \hat{N}_p)\} | \psi_{\text{leads}} \rangle\\
&= \sum_{m=0}^{\infty} \sum_{i_1, \ldots, i_m} \exp\{-\beta \sum_{k=1}^m (\epsilon_{R_{i_k}} - \mu_{p_k} N_{p_k})\} \quad (\text{F.5})\\
&= 1 + \sum_{R_1} e^{-\beta(\epsilon_{R_1} - \mu_{p_1} N_{p_1})} + \sum_{R_1, R_2} e^{-\beta \sum_{k=1}^2 (\epsilon_{R_k} - \mu_{p_k} N_{p_k})} + \ldots\\
&= \prod_{k \sigma p} (1 + e^{-\beta(\epsilon_k - \mu_p)}).
\end{aligned}$$

F. Lead Green's Function

To calculate the Green's function, we have to invert (F.3), which results in (see [86])

$$ig_R^{++} = \frac{1}{1+\rho} \begin{pmatrix} 1 & -\rho(1+i\phi) & -\rho(1+2i\phi) & \cdots \\ 1-i\phi & 1 & -\rho(1+i\phi) & \cdots \\ 1-2i\phi & 1-i\phi & 1 & \ddots \\ \vdots & \vdots & \ddots & \ddots \end{pmatrix}$$
$$+ \mathcal{O}(\delta_t^2), \tag{F.6a}$$

$$ig_R^{+-} = \frac{-\rho}{1+\rho} \begin{pmatrix} \cdots & 1+2i\phi & 1+i\phi & 1 \\ \cdots & 1+i\phi & 1 & 1-i\phi \\ \cdot\cdot\cdot & 1 & 1-i\phi & 1-2i\phi \\ \cdot\cdot\cdot & \cdot\cdot\cdot & \vdots & \vdots \end{pmatrix} + \mathcal{O}(\delta_t^2), \tag{F.6b}$$

$$ig_R^{-+} = \frac{1}{1+\rho} \begin{pmatrix} \vdots & \vdots & \cdot\cdot\cdot & \cdot\cdot\cdot \\ 1-2i\phi & 1-i\phi & 1 & \cdot\cdot\cdot \\ 1-i\phi & 1 & 1+i\phi & \cdots \\ 1 & 1+i\phi & 1+2i\phi & \cdots \end{pmatrix} + \mathcal{O}(\delta_t^2), \text{ and} \tag{F.6c}$$

$$ig_R^{--} = \frac{1}{1+\rho} \begin{pmatrix} \ddots & \ddots & \vdots & \vdots \\ \ddots & 1 & -\rho(1-i\phi) & -\rho(1-2i\phi) \\ \cdots & 1+i\phi & 1 & -\rho(1-i\phi) \\ \cdots & 1+2i\phi & 1+i\phi & 1 \end{pmatrix} + \mathcal{O}(\delta_t^2), \tag{F.6d}$$

where we defined $\rho := \exp\{-\beta(\epsilon_k - \mu_p)\}$. To get from these discrete matrices to the continuous equations (4.59), we first identify $\rho/(1+\rho) = f^+(\epsilon_k - \mu_p)$ and $(1+\rho)^{-1} = f^-(\epsilon_k - \mu_p)$. Second, we notice that any matrix element in row k (corresponding to real-time t) and column l (corresponding to t') comes with a factor $1 \mp i\epsilon_k(k-l)\delta_t/\hbar$, where the sign depends on the *real-time* difference $t - t' \equiv \mp(k-l)\delta_t$. For example, in matrix ig_R^{+-} the *upper right* corner corresponds to the real-time coordinates $t = t' = t_i$

and thus, the matrix element in row 3, column N comes corresponds to real-times $t = t_i + 2\delta_t$ and $t' = t_i$ and comes with a factor $1 - 2i\epsilon_k \delta_t/\hbar$. Then, with the approximation

$$1 - i\epsilon_k(t - t')/\hbar = \exp\{-i\epsilon_k(t - t')/\hbar\} + \mathcal{O}(\delta_t^2), \qquad \text{(F.7)}$$

and (F.2) we obtain the result (4.59) in the limit $\delta_t \to 0$.

Evaluating the γ-matrix

Replacing the summation over the lead wave vectors by the integration over energies ϵ_k in equation (4.61b) yields for $\tilde{t} := t - t' \neq 0$

$$\begin{aligned}
\gamma(p, \tilde{t}) &= \frac{i\Gamma}{\hbar^2} \int \frac{d\epsilon_k}{2\pi} e^{-i\epsilon_k \tilde{t}/\hbar} \times \\
&\quad \cdots \times \begin{pmatrix} \theta(\tilde{t}) f_p^+(\epsilon_k) - \theta(-\tilde{t}) f_p^-(\epsilon_k) & -f_p^+(\epsilon_k) \\ f_p^-(\epsilon_k) & \theta(-\tilde{t}) f_p^+(\epsilon_k) - \theta(\tilde{t}) f_p^-(\epsilon_k) \end{pmatrix} \\
&= \frac{i\Gamma}{2\beta\hbar^2} e^{-i\mu_p \tilde{t}/\hbar} \begin{pmatrix} \theta(\tilde{t}) \hat{f}^+(\tilde{t}) - \theta(-\tilde{t}) \hat{f}^-(\tilde{t}) & -\hat{f}^+(\tilde{t}) \\ \hat{f}^-(\tilde{t}) & \theta(-\tilde{t}) \hat{f}^+(\tilde{t}) - \theta(\tilde{t}) \hat{f}^-(\tilde{t}) \end{pmatrix} \\
&= \frac{\Gamma}{2\beta\hbar^2} \frac{e^{-i\mu_p \tilde{t}/\hbar}}{\sinh[\pi\tilde{t}/(\hbar\beta)]} \begin{pmatrix} -1 & 1 \\ 1 & -1 \end{pmatrix},
\end{aligned}$$
(F.8)

where we calculated the Fourier transform $\hat{f}$ of the Fermi function by a complex contour integration

$$\hat{f}^\pm(\tilde{t}) = \frac{\beta}{\pi} \int d\epsilon_k \, f^\pm(\epsilon_k) e^{-i\epsilon_k \tilde{t}/\hbar} = \frac{\beta}{\pi} \int d\epsilon_k \, f(\epsilon_k) e^{\mp i\epsilon_k \tilde{t}/\hbar}$$

$$= \pm 2i \sum_{m=0}^{\infty} \exp\left\{-(2m+1)\frac{\pi\tilde{t}}{\hbar\beta}\right\} = \pm 2i e^{-\frac{\pi\tilde{t}}{\hbar\beta}} \sum_{m=0}^{\infty} \exp\left\{-\frac{2\pi\tilde{t}}{\hbar\beta}\right\}^m$$

$$= \pm \frac{2ie^{-\frac{\pi \tilde{t}}{\hbar \beta}}}{1 - e^{-\frac{2\pi \tilde{t}}{\hbar \beta}}} = \pm \frac{2i}{e^{\frac{\pi \tilde{t}}{\hbar \beta}} - e^{-\frac{\pi \tilde{t}}{\hbar \beta}}} = \pm i \left[\sinh\left(\frac{\pi \tilde{t}}{\hbar \beta}\right) \right]^{-1}$$
(F.9)

Applying Wick's Theorem

We show how to arrive at the result (4.67) by applying Wick's theorem to $\langle P_K^\uparrow \rangle$. It states that the expectation value of a operator product is equal to the total pairing, i.e., the sum of all possibilities to form contractions, pairs, of creators and annihilators:

$$\begin{aligned}
\langle P_K^\uparrow \rangle &= \langle \hat{d}_\uparrow^\dagger(t_4^<)\hat{d}_\uparrow(t_3^<)\hat{d}_\uparrow^\dagger(t_2)\hat{d}_\uparrow(t_1^<) \rangle + \langle \hat{d}_\uparrow^\dagger(t_4^<)\hat{d}_\uparrow(t_3^<)\hat{d}_\uparrow^\dagger(t_2)\hat{d}_\uparrow(t_1^<) \rangle \\
&= \langle \hat{d}_\uparrow(t_3^<)\hat{d}_\uparrow^\dagger(t_4^<) \rangle \langle \hat{d}_\uparrow(t_1^<)\hat{d}_\uparrow^\dagger(t_2) \rangle - \langle \hat{d}_\uparrow(t_3^<)\hat{d}_\uparrow^\dagger(t_2) \rangle \langle \hat{d}_\uparrow(t_1^<)\hat{d}_\uparrow^\dagger(t_4^<) \rangle \\
&= i^2 [(G_\uparrow^{\text{eff}})^{++}_{t_1^<,t_2} (G_\uparrow^{\text{eff}})^{+-}_{t_3^<,t_4^<} - (G_\uparrow^{\text{eff}})^{+-}_{t_1^<,t_4^<} (G_\uparrow^{\text{eff}})^{++}_{t_3^<,t_2}] \\
&= i^2 \det \begin{pmatrix} (G_\uparrow^{\text{eff}})^{++}_{t_1^<,t_2} & (G_\uparrow^{\text{eff}})^{+-}_{t_1^<,t_4^<} \\ (G_\uparrow^{\text{eff}})^{++}_{t_3^<,t_2} & (G_\uparrow^{\text{eff}})^{+-}_{t_3^<,t_4^<} \end{pmatrix},
\end{aligned}$$
(F.10)

where in the third step the definition of the Green's function was used. This relation could also be obtained in the framework of the fermionic path integal as can be seen in [86], for example. The corresponding expression for $\langle P_K^\downarrow \rangle$ is obtained in exactly the same way.

G

Supplementary Calculations for ISPI

Neglecting Block Connections Beyond the Coherence Time

We use mathematical induction to show that for all $i \geq 2$, the modified diagonal blocks of matrix $\mathbf{X}$ in section 4.2.2 are given by

$$[X]'_{i,i} = [X]_{i,i} - [X]_{i,i-1}[X]^{-1}_{i-1,i-1}[X]_{i-1,i}, \qquad (G.1)$$

when step (4.I2) is applied at each iteration. As this relation is already established for $i = 2$ [see equation (4.73)], we just need to check for the iteration step $i \to i+1$. After iteration $i - 1$, we have

$$\det \mathbf{X} = \det [X]_{1,1} [\prod_{k=2}^{i-1} \det [X]'_{k,k}] \det \begin{pmatrix} [X]'_{i,i} & [X]_{i+1,i} \\ [X]_{i,i+1} & \mathbf{X}_{D-i} \end{pmatrix}. \qquad (G.2)$$

In the i-th iteration cycle, the first step is to eliminate $[X]_{i,i+1}$ in the remaining $D-i+1$ dimensional block matrix. By this, block $[X]_{i+1,i+1}$ is changed to

$$[X]_{i+1,i+1} - [X]_{i+1,i}([X]'_{i,i})^{-1}[X]_{i,i+1}$$
$$= [X]_{i+1,i+1} - [X]_{i+1,i}([X]_{i,i} - [X]_{i,i-1}[X]^{-1}_{i-1,i-1}[X]_{i-1,i})^{-1}[X]_{i,i+1}$$

G. Supplementary Calculations for ISPI

$$= [X]_{i+1,i+1} - [X]_{i+1,i}[X]_{i,i}^{-1}\left(\mathbb{1} - [X]_{i,i-1}[X]_{i-1,i-1}^{-1}[X]_{i-1,i}[X]_{i,i}^{-1}\right)^{-1}[X]_{i,i+1}$$
$$= [X]_{i+1,i+1} - [X]_{i+1,i}[X]_{i,i}^{-1}[X]_{i,i+1}$$
$$- \sum_{k=1}^{\infty} [X]_{i+1,i}[X]_{i,i}^{-1}\left([X]_{i,i-1}[X]_{i-1,i-1}^{-1}[X]_{i-1,i}[X]_{i,i}^{-1}\right)^{k}[X]_{i,i+1}.$$
(G.3)

To this expression, we can apply operation (4.I2). Apparently, all terms in the sum over k, which originate from the higher order expansion orders of the inverse matrix sum in the third line, couple spins in segments $i-1$ and $i+1$. Hence, they can all be neglected. This yields the final result

$$[X]'_{i+1,i+1} = [X]_{i+1,i+1} - [X]_{i+1,i}[X]_{i,i}^{-1}[X]_{i,i+1}.$$
(G.4)

It is important to keep in mind that the whole procedure for the iterative calculation of the determinant rests on a crucial assumption: the absolute value of matrix elements is largest close to the main diagonal and decays rapidly with increasing distance to it. An assumption like (4.I2) would certainly not be justified for an arbitrary block matrix with the same form as X. However, in the spirit of the coherence time approximation, we can argue that non-diagonal blocks of the inverse Green's and Ξ-matrices contribute less to the determinant than diagonal ones. In the expansion (G.3), the first and second term contain zero and two non-diagonal blocks, respectively. All higher orders (in the k-sum) contain four and more and are, hence, considerably smaller. The second term cannot be neglected without loosing all connection between spins in different segments. The higher order terms do not further change the result qualitatively, but merely provide (relatively) small corrections.

G. Supplementary Calculations for ISPI

Expansion of a Block Matrix in Orders of Segment-Connecting Blocks

We expand the inverse of block matrix $\mathbf{X}$ in orders of those matrix blocks $\boxed{X}_{i,i\pm 1}$ that lie on the secondary diagonals and describe connections between spins on different path segments. To do so, we define

$$\mathbf{X}_d = \begin{pmatrix} \boxed{X}_{1,1} & & \\ & \ddots & \\ & & \boxed{X}_{D,D} \end{pmatrix} \text{ and } \mathbf{X}_n = \begin{pmatrix} & \boxed{X}_{1,2} & & \\ \boxed{X}_{2,1} & & \ddots & \\ & \ddots & & \boxed{X}_{D-1,D} \\ & & \boxed{X}_{D,D-1} & \end{pmatrix}, \tag{G.5}$$

so that $\mathbf{X} = \mathbf{X}_d + \mathbf{X}_n$. With this, the inverse of $\mathbf{X}$ can be evaluated to

$$\begin{aligned}
\mathbf{X}^{-1} &= (\mathbf{X}_d + \mathbf{X}_n)^{-1} = \left([\mathbb{1} + \mathbf{X}_n(\mathbf{X}_d)^{-1}]\mathbf{X}_d\right)^{-1} \\
&= (\mathbf{X}_d)^{-1}\left[\mathbb{1} + \mathbf{X}_n(\mathbf{X}_d)^{-1}\right]^{-1} = (\mathbf{X}_d)^{-1} \sum_{k=0}^{\infty}[-\mathbf{X}_n(\mathbf{X}_d)^{-1}]^k \\
&= (\mathbf{X}_d)^{-1} - (\mathbf{X}_d)^{-1}\mathbf{X}_n(\mathbf{X}_d)^{-1} + (\mathbf{X}_d)^{-1}\mathbf{X}_n(\mathbf{X}_d)^{-1}\mathbf{X}_n(\mathbf{X}_d)^{-1} - \dots \\
&=: \mathbf{X}_0^{-1} + \mathbf{X}_1^{-1} + \mathbf{X}_2^{-1} + \dots
\end{aligned} \tag{G.6}$$

In matrix notation, the two lowest order terms are given by

$$\mathbf{X}_0^{-1} = \begin{pmatrix} \boxed{X}_{1,1}^{-1} & & \\ & \ddots & \\ & & \boxed{X}_{D,D}^{-1} \end{pmatrix} \tag{G.7}$$

and

$$\mathbf{X}_1^{-1} = -\begin{pmatrix} & [X]_{1,1}^{-1}[X]_{1,2}[X]_{2,2}^{-1} & & \\ [X]_{2,2}^{-1}[X]_{2,1}[X]_{2,2}^{-1} & & [X]_{2,2}^{-1}[X]_{2,3}[X]_{3,3}^{-1} & \\ & [X]_{3,3}^{-1}[X]_{3,2}[X]_{2,2}^{-1} & & \\ & & & \ddots \end{pmatrix}. \tag{G.8}$$

If we set $(G_\sigma^{\text{eff}})^{-1}$ for $\mathbf{X}$, these terms yield equation (4.75) for the approximate Green's function. To further include the third expansion term from equation (G.6) would contribute the addends

$$\begin{aligned} & [X]_{k,k}^{-1}[X]_{k,k-1}[X]_{k-1,k-1}^{-1}[X]_{k-1,k}[X]_{k,k}^{-1} \\ & + [X]_{k,k}^{-1}[X]_{k,k+1}[X]_{k+1,k+1}^{-1}[X]_{k+1,k}[X]_{k,k}^{-1} \end{aligned} \tag{G.9}$$

to diagonal elements in row k. We can argue in the same way as above, as to why the CTA permits to neglect these contributions. Not only are they of higher order in the non-diagonal blocks and, thus, relatively small. To calculate them also requires knowledge of spins in three consecutive segments ($k, k \pm 1$), which contravenes the CTA and impedes the iterative calculation of $\mathcal{Z}[\eta]$. The second term on the other hand has to be included, since it connects close-lying spins in neighbouring segments.

After the cancellation of the divergent elements in the generating function in section 4.2.3, the situation complicates slightly. The sought-after approximate Green's function is given by the matrix product $D_\sigma^{-1} G_{0,\sigma}^{\text{el}}$. In this case, we approximate the inverse of D_σ according to (4.75) and multiply it with the free Green's matrix in block form. Yet, this straightforward procedure has to be adapted once more to fully comply with the CTA, since diagonal elements would involve spins from both neighbouring segments, while the system only "remembers," at a given iteration, the spins in a path

G. Supplementary Calculations for ISPI

of two segments length. Hence, as far as the system is considered, at iteration $k-1 \to k$ segment $k-1$ is the beginning of the whole "remembered" spin path, while segment k is its end. Therefore, it suffices to assume that the block matrices used for a CTA-conform approximation have block dimension 2. As a result, the approximate value of a matrix product $\mathbf{X}^{-1}\mathbf{Y}$ at iteration step $k-1 \to k$ is given by

$$\begin{pmatrix} \boxed{X}_{k-1,k-1}^{-1} & -\boxed{X}_{k-1,k-1}^{-1}\boxed{X}_{k-1,k}\boxed{X}_{k,k}^{-1} \\ -\boxed{X}_{k,k}^{-1}\boxed{X}_{k,k-1}\boxed{X}_{k-1,k-1}^{-1} & \boxed{X}_{k,k}^{-1} \end{pmatrix} \begin{pmatrix} \boxed{Y}_{k-1,k-1} & \boxed{Y}_{k-1,k} \\ \boxed{Y}_{k,k-1} & \boxed{Y}_{k,k} \end{pmatrix}. \tag{G.10}$$

With $\mathbf{X} = D_\sigma$ and $\mathbf{Y} = G^{\text{el}}_{0,\sigma}$, we obtain equation (4.86).

Inversion of the Inverse Free Electronic Green's Function

To invert $(G^{\text{el}}_{0,\sigma})^{-1}_{t,t'}$ from equation (4.77), we first transform it into Fourier space, where it is finite for all ω, ω' and, hence, can be handled more easily. For this, we replace the time dependent dot-electron fields in the action $S^{\text{el}}_{0,\sigma} = \iint dt\, dt'\, \bar{\eth}_\sigma(t)(G^{\text{el}}_{0,\sigma})^{-1}_{t,t'}\eth_\sigma(t')$ with $\eth_\sigma(t) = (2\pi)^{-1} \int d\omega\, \exp\{-i\omega t\} \times \eth_\sigma(\omega)$ and $\bar{\eth}_\sigma(t) = (2\pi)^{-1} \int d\omega\, \bar{\eth}_\sigma(\omega)\exp\{i\omega t\}$ to obtain

$$\begin{aligned} (G^{\text{el},0}_{\text{dot},\sigma})^{-1}_{\omega,\omega'} &= \begin{pmatrix} 1 & 0 \\ 0 & -1 \end{pmatrix} \iint dt\, dt'\, e^{i\omega t}\delta(t-t')(i\partial_t - \omega^U_\sigma)e^{-i\omega' t'} \\ &= \begin{pmatrix} 1 & 0 \\ 0 & -1 \end{pmatrix} \int dt\, e^{i\omega t}(\omega' - \omega^U_\sigma)e^{-i\omega' t} \quad \text{(G.11)} \\ &= 2\pi\delta(\omega - \omega') \begin{pmatrix} \omega - \omega^U_\sigma & 0 \\ 0 & -\omega + \omega^U_\sigma \end{pmatrix}, \end{aligned}$$

and, with $\gamma(p, t-t') = (2\pi)^{-1} \int d\omega\, \exp\{-i\omega(t-t')\}\gamma(p,\omega)$ [see equation (F.7) in appendix F],

G. Supplementary Calculations for ISPI

$$
\begin{aligned}
(G^0_{\text{env},\sigma})^{-1}_{\omega,\omega'} &= \sum_p \iint dt\, dt'\, e^{i(\omega t - \omega' t')} \gamma(p, t-t') \\
&= \frac{1}{2\pi} \sum_p \iiint dt\, dt'\, d\omega''\, e^{i(\omega-\omega'')t} e^{-i(\omega'-\omega'')t'} \gamma(p, \omega'') \\
&= 2\pi \sum_p \int d\omega''\, \delta(\omega-\omega'')\delta(\omega'-\omega'') \gamma(p, \omega'') \\
&= \frac{2\pi i \Gamma}{\hbar} \delta(\omega-\omega') \sum_p \begin{pmatrix} f_p(\omega) - 1/2 & -f_p(\omega) \\ 1 - f_p(\omega) & f_p(\omega) - 1/2 \end{pmatrix},
\end{aligned}
\qquad \text{(G.12)}
$$

where we used the identity $\int \exp\{\pm i\omega t\}dt = 2\pi\delta(\omega)$ in both derivations. In the last step of (G.12), we exploited the fact that, after replacing $\sum_k$ by an energy integration, the elements of the γ-matrix are essentially given by the Fourier transform of the Fermi function. Strictly speaking, this is only the case for the off-diagonal elements. The diagonal entries contain the Heaviside step functions and require a little more work. Due to the presence of the step function, each addend in a diagonal element of γ leads to a *time-ordered* integral in the third line of equation (G.12). For the $++$ component, we have

$$
\frac{i\Gamma}{2\pi\hbar} \sum_p \iiint dt\, dt'\, d\omega''\, e^{i(\omega-\omega'')t} e^{-i(\omega'-\omega'')t'} \times
$$
$$
\cdots \times [\theta(t-t')f_p^+(\omega) - \theta(t'-t)f_p^-(\omega)] \qquad \text{(G.13)}
$$
$$
= \frac{i\Gamma}{2\pi\hbar} \sum_p \int d\omega'' \Bigg\{ \int_{-\infty}^{\infty} dt \int_{-\infty}^{t} dt'\, e^{i(\omega-\omega'')t} e^{-i(\omega'-\omega'')t'} f_p^+(\omega)
$$
$$
- \int_{-\infty}^{\infty} dt \int_{t}^{\infty} dt'\, e^{i(\omega-\omega'')t} e^{-i(\omega'-\omega'')t'} f_p^-(\omega) \Bigg\}
$$

G. Supplementary Calculations for ISPI

$$
\begin{aligned}
&= \frac{i\Gamma}{2\pi\hbar} \sum_p \int d\omega'' \Big\{ \int_{-\infty}^{\infty} dt\, e^{i(\omega-\omega')t} \int_0^{-\infty} dt'\, e^{i(\omega'-\omega'')t'} f_p^+(\omega) \\
&\quad - \int_{-\infty}^{\infty} dt\, e^{i(\omega-\omega')t} \int_0^{\infty} dt'\, e^{-i(\omega'-\omega'')t'} f_p^-(\omega) \Big\} \\
&= \frac{i\Gamma}{\hbar} \delta(\omega-\omega') \sum_p \int d\omega'' \Big\{ \Big[\pi\delta(\omega'-\omega'') + i\frac{\mathcal{P}}{\omega'-\omega''}\Big] f_p^+(\omega) \\
&\quad - \Big[\pi\delta(\omega'-\omega'') - i\frac{\mathcal{P}}{\omega'-\omega''}\Big] f_p^-(\omega) \Big\} \\
&= \frac{i\Gamma}{\hbar} \delta(\omega-\omega') \sum_p \Big\{ \pi[f_p^+(\omega) - f_p^-(\omega)] \\
&\quad + i\,\mathcal{P} \frac{d\omega''}{\omega'-\omega''}[f_p^+(\omega) + f_p^-(\omega)] \Big\} \\
&= \frac{2\pi i\Gamma}{\hbar} \delta(\omega-\omega') \frac{f_p^+(\omega) - f_p^-(\omega)}{2}, \quad (G.13)
\end{aligned}
$$

where $\mathcal{P}$ denotes the Cauchy principal value. We used the distribution identity $\int_0^\infty \exp\{i\omega t\}\, dt = \pi\delta(\omega) + i\mathcal{P}/\omega$. If we now add equations (G.11) and (G.12) and perform the summation over lead index p, we arrive at the final result (4.2.3) for the free inverse Green's function in frequency space. It can be easily inverted yielding

$$
\begin{aligned}
&(G_{0,\sigma}^{\text{el}})_{\omega,\omega'} \\
&= \frac{2\pi\delta(\omega-\omega')}{(\omega-\omega_\sigma^U)^2 + (\Gamma/\hbar)^2} \times \\
&\quad \times \begin{pmatrix} \omega - \omega_\sigma^U + i\Gamma/\hbar[1-F(\omega)] & -i\Gamma/\hbar F(\omega) \\ i\Gamma/\hbar[2-F(\omega)] & -\omega + \omega_\sigma^U + i\Gamma/\hbar[1-F(\omega)] \end{pmatrix}.
\end{aligned}
$$
(G.14)

The next step of transforming it back into time space is done by complex contour integration. The denominator in front of the matrix ensures that all the integrals converge. Two kinds of integrals appear in the conversion process, which we deal with separately. Using the residue theorem, the first kind evaluates to

$$\int_{-\infty}^{\infty} \frac{d\omega}{2\pi} \frac{(\omega - \omega_\sigma^U) e^{-i\omega(t-t')}}{(\omega - \omega_\sigma^U)^2 + (\Gamma/\hbar)^2} = e^{-i\omega_\sigma^U(t-t')} \int_{-\infty}^{\infty} \frac{d\omega}{2\pi} \frac{\omega \, e^{-i\omega(t-t')}}{\omega^2 + (\Gamma/\hbar)^2}$$

$$= \frac{2\pi i \, e^{-i\omega_\sigma^U(t-t')}}{2\pi} \left[-\theta(t-t') \frac{e^{-\Gamma/\hbar(t-t')}}{2} + \theta(t'-t) \frac{e^{\Gamma/\hbar(t-t')}}{2} \right]$$

$$= -\frac{i \operatorname{sign}(t-t')}{2} e^{-i\omega_\sigma^U(t-t')} e^{-\Gamma/\hbar|t-t'|}.$$

(G.15)

In the second kind of integral, the Fermi function appears in the numerator of the integrand. Therefore, the residue sum also involves the Fermi function's infinitely many residues at the frequencies $i\Omega_m$ with $\Omega_m = (2m+1)\pi/(\hbar\beta)$. Thus, we get

$$\frac{i\Gamma}{\hbar} \int_{-\infty}^{\infty} \frac{d\omega}{2\pi} \frac{f(\omega - \mu_p/\hbar) \, e^{-i\omega(t-t')}}{(\omega - \omega_\sigma^U)^2 + (\Gamma/\hbar)^2}$$

$$= \frac{i\Gamma}{\hbar} \int_{-\infty}^{\infty} \frac{d\omega}{2\pi} \frac{f(\omega) \, e^{-i[\omega + \mu_p/\hbar](t-t')}}{(\omega - \underbrace{[\omega_\sigma^U - \mu_p/\hbar]}_{=:\,\omega_1})^2 + (\Gamma/\hbar)^2}$$

(G.16)

$$= -\frac{\Gamma}{2\pi i \hbar} e^{-i\mu_p/\hbar(t-t')} \int_{-\infty}^{\infty} d\omega \frac{f(\omega) \, e^{-i\omega(t-t')}}{(\omega - \omega_1)^2 + (\Gamma/\hbar)^2}$$

$$= \frac{\Gamma}{\hbar} e^{-i\mu_p/\hbar(t-t')} \Bigg[\theta(t-t')\bigg\{ -\frac{1}{\beta\hbar} \sum_{m=0}^{\infty} \frac{e^{-\Omega_m(t-t')}}{(\omega_1 + i\Omega_m)^2 + (\Gamma/\hbar)^2}$$

$$+ \frac{i\hbar}{2\Gamma} e^{-i[\omega_1 - i\Gamma/\hbar](t-t')} f(\omega_1 - i\Gamma/\hbar) \bigg\}$$

$$+ \theta(t'-t)\bigg\{ \frac{1}{\beta\hbar} \sum_{m=0}^{\infty} \frac{e^{\Omega_m(t-t')}}{(\omega_1 - i\Omega_m)^2 + (\Gamma/\hbar)^2}$$

$$+ \frac{i\hbar}{2\Gamma} e^{-i[\omega_1 + i\Gamma/\hbar](t-t')} f(\omega_1 + i\Gamma/\hbar) \bigg\} \Bigg]$$

G. Supplementary Calculations for ISPI

$$
\begin{aligned}
&= -\frac{\text{sign}(t-t')\,\Gamma}{\beta\hbar^2} e^{-i\mu_p/\hbar(t-t')} e^{-\pi|t-t'|/(\beta\hbar)} \times \\
&\quad \times \sum_{m=0}^{\infty} \frac{\left(e^{-2\pi|t-t'|/(\beta\hbar)}\right)^m}{(\omega_1 + i\,\text{sign}(t-t')\Omega_m)^2 + (\Gamma/\hbar)^2} \\
&\quad + \frac{i}{2} e^{-i\omega_\sigma^U(t-t')} e^{-\Gamma/\hbar|t-t'|} f(\omega_1 - i\,\text{sign}(t-t')\Gamma/\hbar)
\end{aligned}
\tag{G.16}
$$

With these results and the definition

$$
\mathcal{G}_p(t-t') := \frac{\Gamma}{\beta\hbar^2} \sum_{m=0}^{\infty} \frac{e^{-i\mu_p/\hbar(t-t')}\left(e^{-2\pi|t-t'|/(\beta\hbar)}\right)^m}{(\omega_\sigma^U - \mu_p/\hbar + i\,\text{sign}(t-t')\Omega_m)^2 + (\Gamma/\hbar)^2},
\tag{G.17}
$$

we can now write down all four Keldysh components of the free Green's function in real-time space. We obtain $(G_{0,\sigma}^{\text{el}})_{t,t'} := (\tilde{G}_{0,\sigma}^{\text{el}})_{t,t'}[t]^{-2}$ with

$$
(\tilde{G}_{0,\sigma}^{\text{el}})_{t,t'}^{++} = \frac{i}{2} e^{-\Gamma/\hbar|t-t'|} e^{-i\omega_\sigma^U(t-t')} \{f_L^-(\tilde{\omega}_\sigma^U) - f_R^+(\tilde{\omega}_\sigma^U) - \text{sign}(t-t')\}
$$
$$
+ \text{sign}(t-t') e^{-\pi|t-t'|/(\beta\hbar)} \{\mathcal{G}_L(t-t') + \mathcal{G}_R(t-t')\}
\tag{G.18a}
$$

$$
(\tilde{G}_{0,\sigma}^{\text{el}})_{t,t'}^{+-} = (\tilde{G}_{0,\sigma}^{\text{el}})_{t,t'}^{++} - \frac{i}{2} e^{-\Gamma/\hbar|t-t'|} e^{-i\omega_\sigma^U(t-t')}\{1 - \text{sign}(t-t')\}
\tag{G.18b}
$$

$$
(\tilde{G}_{0,\sigma}^{\text{el}})_{t,t'}^{-+} = (\tilde{G}_{0,\sigma}^{\text{el}})_{t,t'}^{++} + \frac{i}{2} e^{-\Gamma/\hbar|t-t'|} e^{-i\omega_\sigma^U(t-t')}\{1 + \text{sign}(t-t')\}
\tag{G.18c}
$$

$$
(\tilde{G}_{0,\sigma}^{\text{el}})_{t,t'}^{--} = (\tilde{G}_{0,\sigma}^{\text{el}})_{t,t'}^{++} + i\,\text{sign}(t-t') e^{-\Gamma/\hbar|t-t'|} e^{-i\omega_\sigma^U(t-t')},
\tag{G.18d}
$$

where $\tilde{\omega}_\sigma^U := \omega_\sigma^U - i\,\text{sign}(t-t')\Gamma/\hbar$ and $[t]$ is a distribution that provides for the correct time dimensions (see below). Apparently, all components explicitly show an exponentially decaying behaviour for growing $|t-t'|$ and are also finite for small up to infinitesimal values of the time difference (see section 4.2.3). At $t = t'$, the imaginary part of $(G_{0,\sigma}^{\text{el}})_{t,t'}$ has a discon-

G. Supplementary Calculations for ISPI

tinuity, which is, however, irrelevant for the cancellation procedure shown in section 4.2.3 (see discussion there).

Units of Continuous Functions and their Discretization

To find a consistent way of attaching time units to continuous versions of discrete matrices such as the inverse Green's function $(G_0^{\text{el}})^{-1}$ in equation (4.28), we consider a matrix $A_{kl} := a_k \delta_{kl} \delta_t$, where the elements a_k have the dimension of time^{-1}. By performing the same steps taken to go from the discrete expression (4.28) to the continuous version (4.31)—namely to pull out one δ_t and, in the limit $\delta_t \to 0$, transform it to the differential dt— we arrive at $A_{t,t'} = a(t)\delta(t - t')$, where $a(t)$ is a suitable function. In the following example

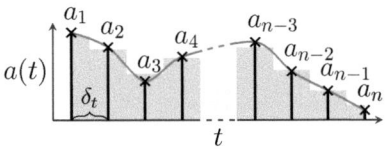

the grey line is a plot of a function $a(t)$ that may be the result of the "continuation" of sequence a_k (black lines with cross marking the value). Obviously, this matrix $A_{t,t'}$ can be re-discretised using equation (4.82). The resulting values $\tilde{a}_k$ are given by the heights of the gray shaded bars, which in turn are fixed by the condition that the area of each bar equals the integral over $a(t)$ over interval $t_k \mp \delta_t/2$. Depending on the size of δ_t and the curve $a(t)$, values a_k and $\tilde{a}_k$ may differ from each other. Within the same accuracy bounds, we can (and will) set $\tilde{a}_k = a(t_k) \equiv a_k$.

Having established that equation (4.82) indeed defines a suitable discretization procedure for matrices like the inverse Green's function, we derive an expression for the unity matrix $\mathbb{1}_{kl} = \delta_{kl}$ that is consistent with it. The obvious ansatz $\mathbb{1}_{t,t'} = \delta(t-t')$ does not have the right dimension, while

G. Supplementary Calculations for ISPI

the straightforward approach that was used above yields $\mathbb{1}_{t,t'} = \delta(t-t')/dt$, which is not a well-formed expression. A better solution is to define $A_{kl} := \mathbb{1}_{kl}\delta_t = \delta_{kl}\delta_t$ to obtain $\mathbb{1}_{t,t'} = \delta(t-t')/\delta_t$, which is well-defined and has the proper dimensions. Unfortunately, this last form depends on the time step—a contradiction to a uniquely defined continuous limit $\delta_t \to 0$. With the key observations that an appropriate unity matrix has to be proportional to $\delta(t-t')$ and an additional factor with dimension time^{-1}, however, we can make the ansatz

$$\mathbb{1}_{t,t'} = \frac{\delta(t-t')}{[t]}, \tag{G.19}$$

where $[t]^{-1}$ can be understood as a function that has the value 1 and carries the dimension of time^{-1}. It is actually a distribution, whose defining property can be derived by plugging (G.19) into (4.82), demanding that this yields back the Kronecker delta.

$$\delta_{kl} \stackrel{!}{=} \int_{t_k-\delta_t/2}^{t_k+\delta_t/2} dt \int_{t_l-\delta_t/2}^{t_l+\delta_t/2} dt' \frac{\delta(t-t')}{[t]} = \delta_{kl} \int_{t_k-\delta_t/2}^{t_k+\delta_t/2} dt \, [t]^{-1} \tag{G.20}$$

This suggests the following general definition of the $[\cdot]$ distribution:

$$\int_{t_a}^{t_b} f(t)[t]^n \, dt := \{F(t_b) - F(t_a)\}(t_b - t_a)^n, \tag{G.21}$$

where n is an integer, $dF(t) = f(t)dt$, and f a test function. From the condition, that $\mathbb{1}^2 = \mathbb{1}$, we can derive the following rule for dimensionally correct multiplication of continuous matrices X and Y (both have to carry the dimension of time^{-2}):

$$(XY)(t,t') = \int dt'' \, X(t,t'')Y(t'',t')[t'']. \tag{G.22}$$

H

Landauer-Büttiker Current and Sequential Flip-Flop Rates

Solving the Rate Equations for the Impurity Spin Orientation

The rate equations (5.1) are inhomogeneous, linear differential equations, which can be solved in a number of ways. Here, we employ the separation of variables after decoupling both equations, using the conservation of probabilities:

$$\dot{P}_\uparrow(t) = \frac{dP_\uparrow}{dt}(t) = W_+ - (W_+ + W_-)P_\uparrow(t) = W_+ - 2\overline{W}P_\uparrow(t) \qquad \text{(H.1)}$$
$$\frac{dP_\downarrow}{dt}(t) = W_- - 2\overline{W}P_\downarrow(t).$$

We are only interested in the time evolution of the spin orientation, given by $\langle\tau_z\rangle(t) = P_\uparrow(t) - P_\downarrow(t)$. Thus, by subtracting the second equation in (H.1) from the first we obtain

$$\frac{d\langle\tau_z\rangle}{dt}(t) = \overbrace{W_+ - W_-}^{w} - 2\overline{W}\langle\tau_z\rangle(t). \qquad \text{(H.2)}$$

The separation of variables yields:

$$\int_{t_i}^{t} dt' = (t - t_i) = \int_{\langle\tau_z\rangle(t_i)}^{\langle\tau_z\rangle(t)} \frac{d\tau_z}{w - 2\overline{W}\tau_z} = -\frac{1}{2\overline{W}} \ln\left[\frac{w - 2\overline{W}\langle\tau_z\rangle(t)}{w - 2\overline{W}\tau_i}\right],$$
$$\text{(H.3)}$$

H. Landauer-Büttiker Current and Sequential Flip-Flop Rates

where we used the short notation $\langle \tau_z \rangle(t_i) = \tau_i$. After applying the exponential function on both sides, this equation can be easily solved for $\langle \tau_z \rangle(t)$ and leads to the solution (5.2).

Calculation of the Landauer-Büttiker Current

To calculate the Landauer-Büttiker current as given by equations (5.5) and (5.6), we have to evaluate basically one kind of integral, whose integrand is the product of a Fermi- and Lorentzian function, which can be done by a complex contour integration. With the fermionic Matsubara frequencies $\Omega_m = (2m+1)\pi/(\hbar\beta)$, we get

$$\int_{-\infty}^{\infty} \frac{f(\omega - \mu/\hbar)\,d\omega}{(\omega - \omega_0)^2 + (\Gamma/\hbar)^2} = \int_{-\infty}^{\infty} \frac{f(\omega)\,d\omega}{(\omega - \underbrace{[\omega_0 - \mu/\hbar]}_{=:\,\tilde{\omega}})^2 + (\Gamma/\hbar)^2}$$

$$= \int_{-\infty}^{\infty} \frac{f(\omega)\,d\omega}{(\omega - \tilde{\omega} + i\Gamma/\hbar)(\omega - \tilde{\omega} - i\Gamma/\hbar)}$$

$$= \pi i \Bigg(\frac{\hbar}{2i\Gamma}\{f(\tilde{\omega} + i\Gamma/\hbar) + f(\tilde{\omega} - i\Gamma/\hbar)\}$$

$$- \frac{1}{\hbar\beta}\sum_{m=0}^{\infty}\bigg\{ \frac{1}{(i\Omega_m - \tilde{\omega} + i\Gamma/\hbar)(i\Omega_m - \tilde{\omega} - i\Gamma/\hbar)}$$

$$- \frac{1}{(i\Omega_m + \tilde{\omega} - i\Gamma/\hbar)(i\Omega_m + \tilde{\omega} + i\Gamma/\hbar)} \bigg\} \Bigg)$$

$$= \frac{\pi\hbar}{\Gamma}\Re f(\tilde{\omega} + i\Gamma/\hbar)$$

$$+ \frac{\pi}{2i\beta\Gamma}\sum_{m=0}^{\infty}\bigg\{ \frac{1}{\Omega_m - i\tilde{\omega} - \Gamma/\hbar} - \frac{1}{\Omega_m - i\tilde{\omega} + \Gamma/\hbar}$$

$$- \frac{1}{\Omega_m + i\tilde{\omega} - \Gamma/\hbar} + \frac{1}{\Omega_m + i\tilde{\omega} + \Gamma/\hbar} \bigg\} = \ldots$$

(H.4)

$$= \frac{\pi\hbar}{\Gamma}\Re f(\tilde{\omega} + i\Gamma/\hbar) \tag{H.4}$$

$$+ \frac{\hbar}{4i\Gamma}\sum_{m=0}^{\infty}\left\{\frac{1}{m+1/2-\frac{i\beta}{2\pi}(\hbar\tilde{\omega}-i\Gamma)} - \frac{1}{m+1/2-\frac{i\beta}{2\pi}(\hbar\tilde{\omega}+i\Gamma)}\right.$$

$$\left. - \frac{1}{m+1/2+\frac{i\beta}{2\pi}(\hbar\tilde{\omega}+i\Gamma)} + \frac{1}{m+1/2+\frac{i\beta}{2\pi}(\hbar\tilde{\omega}-i\Gamma)}\right\}$$

$$= \frac{\pi\hbar}{\Gamma}\Re f(\zeta/\hbar) + \frac{\hbar}{4i\Gamma}\left\{\Psi^{(0)}\left(\frac{1}{2}+\frac{i\beta}{2\pi}\zeta\right) - \Psi^{(0)}\left(\frac{1}{2}-\frac{i\beta}{2\pi}\zeta^*\right)\right.$$

$$\left. - \Psi^{(0)}\left(\frac{1}{2}+\frac{i\beta}{2\pi}\zeta^*\right) + \Psi^{(0)}\left(\frac{1}{2}-\frac{i\beta}{2\pi}\zeta\right)\right\}$$

$$= \frac{\hbar}{\Gamma}\left\{\pi\Re f(\zeta/\hbar) + \frac{1}{2}\Im\Psi^{(0)}\left(\frac{1}{2}+\frac{i\beta}{2\pi}\zeta\right) - \frac{1}{2}\Im\Psi^{(0)}\left(\frac{1}{2}+\frac{i\beta}{2\pi}\zeta^*\right)\right\},$$

where we defined $\zeta := \hbar\omega_0 - \mu + i\Gamma$ and used the following identity for the difference of digamma functions $\Psi^{(0)}(z)$:

$$\Psi^{(0)}(z_1) - \Psi^{(0)}(z_2) = \sum_{m=0}^{\infty}\left(\frac{1}{m+z_2} - \frac{1}{m+z_1}\right). \tag{H.5}$$

By setting $\omega_0 = \omega_\sigma^U$ in (H.4) and plug it into the expression for the LB current, we obtain (5.8).

Calculating the Sequential Flip-Flop Rates

To arrive at the kinetic equation (5.15) for the flip-flop probability, we first plug in the r.h.s. of (5.14) into (5.9). This yields

$$P(-\tau, \Delta_t | \tau, 0)$$
$$= (\Pi_0 \Sigma \Pi)_{-\tau \leftarrow \tau}^{-\tau \leftarrow \tau}(\Delta_t, 0) \tag{H.6}$$
$$= \sum_{\tau_1,\ldots,\tau_4}\int_0^{\Delta_t}dt\int_0^t dt'\,(\Pi_0)_{-\tau\leftarrow\tau_3}^{-\tau\leftarrow\tau_4}(\Delta_t,t)\,\Sigma_{\tau_3\leftarrow\tau_1}^{\tau_4\leftarrow\tau_2}(t,t')\,\Pi_{\tau_1\leftarrow\tau}^{\tau_2\leftarrow\tau}(t',0) = \ldots$$

$$= \sum_{\tau_1} \int_0^{\Delta_t} dt \int_0^t dt'\, (\Pi_0)_{-\tau \leftarrow -\tau}^{-\tau \leftarrow -\tau}(\Delta_t, t)\, \Sigma_{-\tau \leftarrow \tau_1}^{-\tau \leftarrow \tau_1}(t, t')\, P(\tau_1, t'|\tau, 0). \tag{H.6}$$

In the last step, we used the fact that due to the electron number conservation the real-time late state indices of a diagram have to be equal, if its early indices are. This allows to identify the full propagator matrix element in the integral with the conditional probability to find the impurity in state $|\tau_1\rangle$ at time t', when it was in state $|\tau\rangle$ at time 0. The result is derived with respect to Δ_t:

$$\frac{dP}{d\Delta_t}(-\tau, \Delta_t|\tau, 0)$$

$$= \sum_{\tau_1} \Big\{ \frac{1}{d\Delta_t} \int_{\Delta_t - d\Delta_t}^{\Delta_t} dt \int_0^t dt'\, (\Pi_0)_{-\tau \leftarrow -\tau}^{-\tau \leftarrow -\tau}(\Delta_t, t)\, \Sigma_{-\tau \leftarrow \tau_1}^{-\tau \leftarrow \tau_1}(t, t')\, P(\tau_1, t'|\tau, 0)$$

$$+ \int_0^{\Delta_t} dt \int_0^t dt'\, \left(\frac{d\Pi_0}{d\Delta_t}\right)_{-\tau \leftarrow -\tau}^{-\tau \leftarrow -\tau}(\Delta_t, t)\, \Sigma_{-\tau \leftarrow \tau_1}^{-\tau \leftarrow \tau_1}(t, t')\, P(\tau_1, t'|\tau, 0) \Big\}$$

$$= \sum_{\tau_1} \int_0^t dt'\, \Sigma_{-\tau \leftarrow \tau_1}^{-\tau \leftarrow \tau_1}(\Delta_t, t')\, P(\tau_1, t'|\tau, 0), \tag{H.7}$$

where we used $(\Pi_0)(\Delta_t, \Delta_t) = \hat{\mathbb{1}}$ and

$$\left(\frac{d\Pi_0}{d\Delta_t}\right)_{-\tau \leftarrow -\tau}^{-\tau \leftarrow -\tau}(\Delta_t, t)$$

$$= \frac{d}{d\Delta_t} \mathrm{Tr}_{\mathrm{el}}\{\hat{\rho}_{\mathrm{el}}\langle\tau|\hat{U}_0(0, \Delta_t)|-\tau\rangle\langle-\tau|\hat{U}_0(\Delta_t, 0)|\tau\rangle\}$$

$$= \mathrm{Tr}_{\mathrm{el}}\{\hat{\rho}_{\mathrm{el}}\langle\tau|\partial_{\Delta_t}\hat{U}_0(0, \Delta_t)|-\tau\rangle\langle-\tau|\hat{U}_0(\Delta_t, 0)|\tau\rangle\}$$

$$+ \mathrm{Tr}_{\mathrm{el}}\{\hat{\rho}_{\mathrm{el}}\langle\tau|\hat{U}_0(0, \Delta_t)|-\tau\rangle\langle-\tau|\partial_{\Delta_t}\hat{U}_0(\Delta_t, 0)|\tau\rangle\} \tag{H.8}$$

$$= \frac{i}{\hbar} \mathrm{Tr}_{\mathrm{el}}\{\hat{\rho}_{\mathrm{el}}\langle\tau|\hat{U}_0(0, \Delta_t)\hat{H}_0|-\tau\rangle\langle-\tau|\hat{U}_0(\Delta_t, 0)|\tau\rangle\}$$

$$- \frac{i}{\hbar} \mathrm{Tr}_{\mathrm{el}}\{\hat{\rho}_{\mathrm{el}}\langle\tau|\hat{U}_0(0, \Delta_t)|-\tau\rangle\langle-\tau|\hat{H}_0\hat{U}_0(\Delta_t, 0)|\tau\rangle\} = 0.$$

H. Landauer-Büttiker Current and Sequential Flip-Flop Rates

The derivative vanishes, since the Hamiltonian $\hat{H}_0 \equiv \hat{H} - \hat{H}_{\text{int}}^{\perp}$ commutes both with its corresponding time evolution operator $\hat{U}_0$ and the projector $|\tau\rangle\langle\tau|$. The subscript 'el' refers to the whole electronic subsystem (both the lead and dot electrons). The sequential diagrams in equation (5.17) that give the sequential, second-order flip-flop rates evaluate to:[1]

$$\begin{aligned}
W_{-\tau}^{(2)} &= -\frac{J^2}{4\hbar^2} \int_0^\infty dt \Big\{ \langle \bar{d}_{-\tau}^-(0) d_\tau^-(0) \bar{d}_\tau^+(t) d_{-\tau}^+(t) \rangle_\zeta \\
&\quad + \langle \bar{d}_{-\tau}^-(t) d_\tau^-(t) \bar{d}_\tau^+(0) d_{-\tau}^+(0) \rangle_\zeta \Big\} \Big|_{\{\zeta\}=0} \\
&= \frac{J^2}{4\hbar^2} \int_0^\infty dt \Big\{ (G_{0,-\tau}^{\text{el}})^{+-}(t) (G_{0,\tau}^{\text{el}})^{-+}(-t) + (G_{0,-\tau}^{\text{el}})^{+-}(-t) (G_{0,\tau}^{\text{el}})^{-+}(t) \Big\} \\
&= \frac{J^2}{16\pi^2\hbar^2} \int_0^\infty dt \int_{-\infty}^\infty d\omega \int_{-\infty}^\infty d\omega' \Big\{ (G_{0,-\tau}^{\text{el}})^{+-}(\omega)(G_{0,\tau}^{\text{el}})^{-+}(\omega') e^{-i(\omega-\omega')t} \\
&\quad + (G_{0,-\tau}^{\text{el}})^{+-}(\omega)(G_{0,\tau}^{\text{el}})^{-+}(\omega') e^{i(\omega-\omega')t} \Big\} \\
&= \frac{J^2}{16\pi^2\hbar^2} \int_{-\infty}^\infty d\omega \int_{-\infty}^\infty d\omega' (G_{0,-\tau}^{\text{el}})^{+-}(\omega)(G_{0,\tau}^{\text{el}})^{-+}(\omega') \int_{-\infty}^\infty e^{-i(\omega-\omega')t} dt \\
&= \frac{J^2}{8\pi\hbar^2} \int_{-\infty}^\infty d\omega \, (G_{0,-\tau}^{\text{el}})^{+-}(\omega)(G_{0,\tau}^{\text{el}})^{-+}(\omega),
\end{aligned}$$

(H.9)

where the expectation value with subscript 'ζ' is taken from equation (5.3) and the we used the free Green's function from equation (G.14). Now, by going from $(G_{0,\sigma}^{\text{el}})$ to the phenomenological Green's function $(G_{0,\sigma}^{\text{el},J})$ from equation (5.7), we arrive at the final expression (5.18) for the inverse relaxation time in case of sequential flip-flops.

---

[1] This last derivation and the short paragraph at the end were slightly changed to account for the corresponding modification in section 5.3 on page 192. The final result is unchanged.

Danksagung

An dieser Stelle möchte ich den vielen Menschen danken, die dazu beigetragen haben, dass diese Arbeit entstehen konnte. An erster Stelle gilt mein Dank natürlich Frau Prof. Dr. Daniela Pfannkuche – dafür, dass sie mir die Möglichkeit eröffnete, in ihrer tollen Gruppe zu arbeiten, für das große Vertrauen, dass sie mir immer entgegen brachte, ihre exzellente fachliche Anleitung und nicht zuletzt die überaus angenehme und heitere Arbeitsatmosphäre. Ihre große Intuition, Energie und der "Blick für das Ganze" waren nicht nur immer eine große Hilfe, sondern auch eine Inspiration für mich.

Besonderer Dank gebührt auch Prof. Dr. Michael Thorwart – ein Dank, der weit über die Übernahme des Zeitgutachtens hinausgeht. Trotz der meist großen räumlichen Entfernung zwischen uns war er mir immer eine große Hilfe und wie ein zweiter Betreuer und Mentor. Die Arbeit mit ihm war vom ersten Tag an sehr freundschaftlich geprägt und neben den fachlichen Gesprächen, die wesentlich zum Gelingen dieser Arbeit beitrugen, diskutierte er mit mir auch "Karrierefragen" und gab mir ganz allgemein das Gefühl, ein Mitglied seiner Gruppe zu sein (inklusive einiger sehr gemütlicher Abende im Kreise seiner Kollegen und Familie).

Herzlichen Dank auch an Prof. Dr. Jürgen König dafür, dass er freundlicherweise das Zweitgutachtens bei meiner Disputation anfertigte, wofür er in aller Frühe aus Diusburg anreisen musste. Herrn Dr. Georg Steinbrück danke ich, dass er sich so kurzfristig bereiterklärt hat, den Vorsitz des Prüfungsausschusses zu übernehmen. Daniela und Michael haben außerdem viel Zeit und Energie in die Korrektur meiner Doktorarbeit investiert. Es versteht sich von selbst, dass sie in vorliegender Form ohne diese Hilfe nicht entstanden wäre. Vielen Dank für die Mühe!

Stephan Weiß und Jens Eckel möchte ich danken für die große Hilfe bei der Einarbeitung in die ISPI-Methode während meiner vier Wochen in Düsseldorf. Aber auch für die Hilfe, Diskussionen und Zusammenarbeit in den Jahren darauf. Jens und Stephan teilten auch ihre Erfahrung mit Numerik und Programm-Optimierung mit mir – unabdingbar, um den Rechenaufwand meines Programms in Grenzen zu halten. Vielen Dank auch für die wertvollen Korrekturen zu diversen Postern, Abstrakts und Konferenzbeiträgen. Darüber hinaus möchte ich Herr Prof. Dr. Reinhold Egger für seine Hilfe danken und dafür, dass er den Fortgang meiner Arbeit immer aufmerksam verfolgte.

Bernd Güde danke ich nicht nur dafür, dass er seine gute Laune, seinen Optimismus und seine Mittagszeit so oft mit mir geteilt hat; er hat sich auch unmittelbar vor und in der ganzen Zeit nach der Einreichung der Arbeit bis zum glücklichen Ende des Verfahrens immer Zeit genommen, mich durch das Dickicht der Behördengänge und Verfahrensschritte zu führen, die Arbeit nach "Druckfehlern" durchzusehen, mit seiner Schlüsselgewalt für Musik bei der Disputationsfeier zu sorgen, u.v.m. Ich danke Benjamin Baxevanis für die vielen nützlichen Diskussionen über Physik und Programmentwicklung, sein großes Interesse an meiner Arbeit und die sorgsame Reiseplanung für diverse Konferenzen. Bei Frank Deuretzbacher möchte ich mich für dafür bedanken, dass er mir einen Einblick in das Gebiet der kalten Quantengase und damit die Gelegenheit ermöglichte, zu einem sehr schönen Papier beizutragen. Alexander Lieder und Hosnie Safaei-Katoli danke ich für die ausgesprochen angenehmen Jahre, die sie mit mir das Büro geteilt haben. Alexander hat mir durch seine Fragen oft genug die Lücken in meinem Wissen aufgezeigt und durch seinen Rat in sportlichen Dingen auch über das Berufliche hinaus einen schweren Eindruck in meinem Leben hinterlassen (Kettlebell). Hosnie hat durch ihre sanfte, freundliche Art stets für eine heitere Atmosphäre gesorgt. Frank Hellmuth danke ich für einige sehr schöne Abende bei Sushi und Wein, ausgedehnte persönliche Gespräche bis zum Morgengrauen und dafür, dass er mich immer so nachdrücklich ermutigt hat, der Wissenschaft treu zu bleiben.

Insgesamt bin sehr dankbar für die schöne Zeit, die ich in der Gruppe der "Quantentheorie der kondensierten Materie" verbringen durfte. Die Möglichkeit zur Zusammenarbeit mit so vielen interessanten, sympathischen und hilfsbereiten Menschen habe ich immer als Privileg empfunden. Euch allen danke ich für die wunderbare Arbeitsatmosphäre und die unzähligen interessanten Gespräche über Physik, aber auch über all das, was nicht Physik ist, sowie das Persönliche, das ich mit Euch teilen durfte.

Ich kann hier nicht jedem für alles danken, aber zumindest danke ich: Stellan Bohlens für Rat und Hilfe, vor allem in Bewerbungsfragen, und viele anregende Gespräche; Alexander Chudnovskyi dafür, dass er mich immer an seiner großen Fachkenntnis teilhaben ließ, wenn ich ihn darum bat; Christoph Hübner für angenehme Gespräche und das lebensrettende Franzbrötchen im entscheidenden Augenblick; Philipp Knake für seine unzähligen Anekdoten (auch und gerade aus der algebraischen Quantenfeldtheorie); Benjamin Krüger für die intensiven Gespräche, auch zu persönlichen Themen; Tim Ludwig für seine unnachahmliche, lustige Art und gnadenlose Direktheit; Peter Moraczewski für den intensiven, häufig philosophischen Gedankenaustausch und seinen unerschöpflichen Vorrat an Lebensweisheiten; Holger Niehus für die anregenden Diskussionen über alles vom Fahrrad bis zur Differentialgeometrie; Eva-Maria Richter, die mit ihrer fröhlichen, lebenslustigen und charmanten Art stets für gute Stimmung gesorgt und auch die Federführung bei der Konstruktion meines Doktorhuts übernommen hat; Dirk-Sören Lühmann für seinen Rat und ein offenes Ohr für meine Fragen; Jacek Swiebodzinski für seine Weisheit und den hintergründigen Humor.

Natürlich möchte ich auch ganz herzlich meiner Familie und meinen Freunden danken. Sie gaben mir den Rückhalt und das Vertrauen, dass ich brauchte, um diese Arbeit fertigzustellen und unterstützten mich in allem, was ich tat. Meine Eltern Helmut und Ilona haben mir in der entscheidenden Phase ganz entscheidend den Rücken freigehalten und mir geholfen, wo sie nur konnten. Das gilt auch für meinen Bruder Etienne, meine Schwester Denise und die ganze Familie Reck. Besonders dankbar bin dem kleinen Sonnenschein Emma, die mit ihrem strahlenden Lächeln so manchen tristen Arbeitstag erhellte. Außerdem danke ich noch meinen guten Freunden Aleksander, Christoph, Franz, Jan und Moritz für die viele Ablenkung von der Physik, die so wichtig für das Gelingen einer solchen Arbeit ist.

Bibliography

[1] W. Heisenberg. Über quantentheoretische Umdeutung kinematischer und mechanischer Beziehungen. *Zeitschrift für Physik A Hadrons and Nuclei*, 33:879, 1925. 11

[2] M. Born and P. Jordan. Zur Quantenmechanik. *Zeitschrift für Physik A Hadrons and Nuclei*, 34:858, 1925.
M. Born, W. Heisenberg, and P. Jordan. Zur Quantenmechanik II. *Zeitschrift für Physik A Hadrons and Nuclei*, 35:557, 1926.

[3] E. Schrödinger. Quantisierung als Eigenwertproblem (Teil 1 - 4). *Annalen der Physik*, 1926. 11

[4] Louis de Broglie. *Recherches sur la théorie des quanta.* Janvier-Fevrier, 1925. 11

[5] M. Planck. Zur Theorie des Gesetzes der Energieverteilung im Normalspektrum. *Verhandlungen der Deutschen physikalischen Gesellschaft*, 2:237, 1900. 11

[6] A. Einstein. Über einen die Erzeugung und Verwandlung des Lichtes betreffenden heuristischen Gesichtspunkt. *Annalen der Physik*, 322:132, 1905. 11

[7] Niels Bohr. On the constitution of atoms and molecules. *Philosophical Magazine*, 1913. 12

[8] Robert S. Mulliken. Electronic structures of polyatomic molecules and valence. II. General considerations. *Physical Review*, 41:49, 1932. 12

[9] J. A. Wheeler and W. H. Zurek, editors. *Quantum Theory and Measurement*. Princeton University Press, first edition, 1983. 12

[10] J. v. Neumann. Beweis des Ergodensatzes und des H-Theorems in der neuen Mechanik. *Zeitschrift für Physik A Hadrons and Nuclei*, 57:30, 1929.

[11] W. Pauli and M. Fierz. Über das H-Theorem in der Quantenmechanik. *Zeitschrift fur Physik*, 106:572, 1937.

[12] E. Joos, H. D. Zeh, C. Kiefer, D. Giulini, J. Kupsch, and I.-O. Stamatescu. *Decoherence and the Appearance of a Classical World in Quantum Theory*. Springer Berlin, second edition, 2003. 12

[13] Yoseph Imry. *Introduction to Mesoscopic Physics*. Oxford University Press, first edition, 1997. 12

[14] R. P. Feynman and F. L. Vernon. The theory of a general quantum system interacting with a linear dissipative system. *Annals of Physics*, 24:118, 1963. 12

[15] A. O. Caldeira and A. J. Leggett. Influence of dissipation on quantum tunneling in macroscopic systems. *Physical Review Letters*, 46:211, 1981.

[16] A. O. Caldeira and A. J. Leggett. Path integral approach to quantum Brownian motion. *Physica A: Statistical and Theoretical Physics*, 121:587, 1983.

[17] A. O. Caldeira and A. J. Leggett. Quantum tunnelling in a dissipative system. *Annals of Physics*, 149:374, 1983.

[18] Th. Dittrich, p. Hänggi, G.-L. Ingold, B. Kramer, G. Schön, and W. Zwerger. *Quantum Transport and Dissipation*. Wiley-VCH Weinheim, first edition, 1998.

[19] Ulrich Weiss. *Quantum dissipative systems*. World Scientific Singapore, second edition, 1999. 12

[20] P.M. Zeitzoff and J.E. Chung. A perspective from the 2003 ITRS: MOSFET scaling trends, challenges, and potential solutions. *Circuits and Devices Magazine, IEEE*, 21:4, 2005. ISSN 8755-3996. 12

[21] Tapash Chakraborty. *Quantum Dots: A Survey Of The Properties Of Artificial Atoms*. North Holland, first edition, 1999. 12

[22] Stephanie M. Reimann and Matti Manninen. Electronic structure of quantum dots. *Reviews of Modern Physics*, 74:1283, 2002.

[23] Y. Masumoto and T. Takagahara, editors. *Semiconductor Quantum Dots*. Springer-Verlag, Berlin, 2002.

[24] Zhiming M. Wang. *Self-assembled Quantum Dots*. Springer, New York, 2007. 13

[25] R. Hanson, L. P. Kouwenhoven, J. R. Petta, S. Tarucha, and L. M. K. Vandersypen. Spins in few-electron quantum dots. *Reviews of Modern Physics*, 79:1217, 2007. 12, 13, 73

[26] Michael Galperin, Mark A Ratner, and Abraham Nitzan. Molecular

transport junctions: vibrational effects. *Journal of Physics: Condensed Matter*, 19:103201, 2007. 12

[27] Sense Jan van der Molen and Peter Liljeroth. Charge transport through molecular switches. *Journal of Physics: Condensed Matter*, 22:133001, 2010. 12

[28] Shahal Ilani and Paul L. McEuen. Electron transport in carbon nanotubes. *Annual Review of Condensed Matter Physics*, 1:1, 2010. 13

[29] D. S. L. Abergel, V. Apalkov, J. Berashevich, K. Ziegler, and Tapash Chakraborty. Properties of graphene: a theoretical perspective. *Advances in Physics*, 59:261, 2010. 13

[30] J. Leuthold, C. Koos, and W. Freude. Nonlinear silicon photonics. *Nature Photonics*, 4:535, 2010. 13

[31] L. P. Kouwenhoven, C. M. Markus, P. L. McEuen, S. Tarucha, R. M. Westervelt, and N. S. Wingreen. *Mesoscopic Electron Transport, edited by L. L. Sohn, L. P. Kouwenhoven, and Gerd Schön*. Kluwer Academic Publishers, Dordrecht, Boston, London, 1997. 13, 14, 21

[32] L P Kouwenhoven, D G Austing, and S Tarucha. Few-electron quantum dots. *Reports on Progress in Physics*, 64:701, 2001. 13, 14, 16, 17, 62, 69, 70

[33] Mikael T. Björk, Claes Thelander, Adam E. Hansen, Linus E. Jensen, Magnus W. Larsson, L. Reine Wallenberg, and Lars Samuelson. Few-electron quantum dots in nanowires. *Nano Letters*, 4:1621, 2004. 13

[34] C. Fasth, A. Fuhrer, L. Samuelson, Vitaly N. Golovach, and Daniel Loss. Direct measurement of the spin-orbit interaction in a two-

electron InAs nanowire quantum dot. *Physical Review Letters*, 98: 266801, 2007. 73

[35] A. Pfund, I. Shorubalko, K. Ensslin, and R. Leturcq. Suppression of spin relaxation in an InAs nanowire double quantum dot. *Physical Review Letters*, 99:036801, 2007. 13, 73

[36] David L. Klein, Paul L. McEuen, Janet E. Bowen Katari, Richard Roth, and A. Paul Alivisatos. An approach to electrical studies of single nanocrystals. *Applied Physics Letters*, 68:2574, 1996. 13

[37] Dmitri V. Talapin, Jong-Soo Lee, Maksym V. Kovalenko, and Elena V. Shevchenko. Prospects of colloidal nanocrystals for electronic and optoelectronic applications. *Chemical Reviews*, 110:389, 2010. 13

[38] A. L. Rogach, editor. *Semiconductor Nanocrystal Quantum Dots*. Springer Wien, first edition, 2008.

[39] S. Guéron, Mandar M. Deshmukh, E. B. Myers, and D. C. Ralph. Tunneling via individual electronic states in ferromagnetic nanoparticles. *Physical Review Letters*, 83:4148, 1999.

[40] J. R. Petta and D. C. Ralph. Studies of spin-orbit scattering in noble-metal nanoparticles using energy-level tunneling spectroscopy. *Physical Review Letters*, 87:266801, 2001.

[41] D. C. Ralph, C. T. Black, and M. Tinkham. Spectroscopic measurements of discrete electronic states in single metal particles. *Physical Review Letters*, 74:3241, 1995. 13

[42] Patrik Recher and Björn Trauzettel. Quantum dots and spin qubits in graphene. *Nanotechnology*, 21:302001, 2010. 13

[43] Cees Dekker. Carbon nanotubes as molecular quantum wires. *Physics Today*, 52:22, 1999. 13

[44] Sami Sapmaz, Pablo Jarillo-Herrero, Leo P Kouwenhoven, and Herre S J van der Zant. Quantum dots in carbon nanotubes. *Semiconductor Science and Technology*, 21:S52, 2006. 13

[45] Jiwoong Park, Abhay N. Pasupathy, Jonas I. Goldsmith, Connie Chang, Yuval Yaish, Jason R. Petta, Marie Rinkoski, James P. Sethna, Hector D. Abruna, Paul L. McEuen, and Daniel C. Ralph. Coulomb blockade and the Kondo effect in single-atom transistors. *Nature*, 417:722, 2002. 13

[46] Herbert Schoeller and Gerd Schön. Mesoscopic quantum transport: Resonant tunneling in the presence of a strong Coulomb interaction. *Physical Review B*, 50:18436, 1994. 17, 21, 26, 34, 53, 62, 74, 75, 190, 212, 215

[47] Vitaly N. Golovach and Daniel Loss. Transport through a double quantum dot in the sequential tunneling and cotunneling regimes. *Physical Review B*, 69:245327, 2004. 14, 18, 71, 77

[48] Dietmar Weinmann, Wolfgang Häusler, and Bernhard Kramer. Spin blockades in linear and nonlinear transport through quantum dots. *Physical Review Letters*, 74:984, 1995.

[49] B. Baxevanis, D. Becker, J. Gutjahr, P. Moraczewski, and D. Pfannkuche. *Quantum Materials, Lateral Semiconductor Nanostructures, Hybrid Systems and Nanocrystals*, chapter The Different Faces of Coulomb Interaction in Transport through Quantum Dot Systems. Springer, 2010. 13

Bibliography

[50] Alexander V. Khaetskii and Yuli V. Nazarov. Spin relaxation in semiconductor quantum dots. *Physical Review B*, 61:12639, 2000. 13, 16, 17, 69, 70

[51] R. Oulton, A. Greilich, S. Yu. Verbin, R. V. Cherbunin, T. Auer, D. R. Yakovlev, M. Bayer, I. A. Merkulov, V. Stavarache, D. Reuter, and A. D. Wieck. Subsecond spin relaxation times in quantum dots at zero applied magnetic field due to a strong electron-nuclear interaction. *Physical Review Letters*, 98:107401, 2007.

[52] Michael Thorwart. *Tunneling and vibrational relaxation in driven multilevel systems*. PhD thesis, Universität Augsburg, 2000. 164, 168

[53] Vitaly N. Golovach, Alexander Khaetskii, and Daniel Loss. Phonon-induced decay of the electron spin in quantum dots. *Physical Review Letters*, 93:016601, 2004. 16, 17, 70, 85

[54] Axel Thielmann, Matthias H. Hettler, Jürgen König, and Gerd Schön. Super-Poissonian noise, negative differential conductance, and relaxation effects in transport through molecules, quantum dots, and nanotubes. *Physical Review B*, 71:045341, 2005. 74, 76

[55] Ireneusz Weymann and Jozef Barnaś. Effect of intrinsic spin relaxation on the spin-dependent cotunneling transport through quantum dots. *Physical Review B*, 73:205309, 2006. 17, 70, 71, 82

[56] Peter Stano and Jaroslav Fabian. Orbital and spin relaxation in single and coupled quantum dots. *Physical Review B*, 74:045320, 2006.

[57] R. Hanson, B. Witkamp, L. M. K. Vandersypen, L. H. Willems van Beveren, J. M. Elzerman, and L. P. Kouwenhoven. Zeeman energy

and spin relaxation in a one-electron quantum dot. *Physical Review Letters*, 91:196802, 2003. 16, 69

[58] W. A. Coish, V. N. Golovach, J. C. Egues, and D. Loss. Measurement, control, and decay of quantum-dot spins. *physica status solidi (b)*, 243:3658, 2006. 16, 17, 70

[59] J. Voss and D. Pfannkuche. Electron spin relaxation in GaAs quantum dot systems - The role of the hyperfine interaction. *ArXiv e-prints*, 0712.2376, 2007. 17, 70

[60] W. Yang and K. Chang. Spin relaxation in diluted magnetic semiconductor quantum dots. *Physical Review B*, 72:075303, 2005. 13

[61] M. A. Kastner. The single-electron transistor. *Reviews of Modern Physics*, 64:849, 1992. 14, 62

[62] Jürgen König. *Quantum Fluctuations in the Single-Electron Transistor*. Shaker Verlag, Aachen, 1999. 78

[63] M. Leijnse and M. R. Wegewijs. Kinetic equations for transport through single-molecule transistors. *Physical Review B*, 78:235424, 2008.

[64] Conrad R. Wolf, Klaus Thonke, and Rolf Sauer. Single-electron transistors based on self-assembled silicon-on-insulator quantum dots. *Applied Physics Letters*, 96:142108, 2010. 14

[65] S. A. Wolf, D. D. Awschalom, R. A. Buhrman, J. M. Daughton, S. von Molnár, M. L. Roukes, A. Y. Chtchelkanova, and D. M. Treger. Spintronics: A Spin-Based Electronics Vision for the Future. *Science*, 294:1488, 2001. 14, 69

[66] Igor Žutić, Jaroslav Fabian, and S. Das Sarma. Spintronics: Fundamentals and applications. *Reviews of Modern Physics*, 76:323, 2004.

[67] Ian Appelbaum, Biqin Huang, and Douwe J. Monsma. Electronic measurement and control of spin transport in silicon. *Nature*, 447: 295, 2007.

[68] D. D. Awschalom and M. E. Flatté. Challenges for semiconductor spintronics. *Nature Physics*, 3:153, 2007. 28, 69

[69] S.D. Bader and S.S.P. Parkin. Spintronics. *Annual Review of Condensed Matter Physics*, 1:71, 2010. 14

[70] Patrik Recher, Eugene V. Sukhorukov, and Daniel Loss. Quantum dot as spin filter and spin memory. *Physical Review Letters*, 85:1962, 2000. 14, 15

[71] Al. L. Efros, E. I. Rashba, and M. Rosen. Paramagnetic ion-doped nanocrystal as a voltage-controlled spin filter. *Physical Review Letters*, 87:206601, 2001. 19

[72] J. Klinovaja, M. J. Schmidt, B. Braunecker, and D. Loss. Helical modes in carbon nanotubes generated by strong electric fields. *ArXiv e-prints*, 1011.3630, 2010. 14

[73] David P. DiVincenzo. Quantum computation. *Science*, 270:255, 1995. 15

[74] Andrew Steane. Quantum computing. *Reports on Progress in Physics*, 61:117, 1998.

[75] Charles H. Bennett and David P. DiVincenzo. Quantum information and computation. *Nature*, 404:247, 2000. 15

[76] John J. L. Morton, Alexei M. Tyryshkin, Richard M. Brown, Shyam Shankar, Brendon W. Lovett, Arzhang Ardavan, Thomas Schenkel, Eugene E. Haller, Joel W. Ager, and S. A. Lyon. Solid-state quantum memory using the 31P nuclear spin. *Nature*, 455:1085, 2008. 15

[77] Yuriy Makhlin, Gerd Scohn, and Alexander Shnirman. Josephson-junction qubits with controlled couplings. *Nature*, 398:305, 1999. 15

[78] Y. Nakamura, Yu A. Pashkin, and J. S. Tsai. Coherent control of macroscopic quantum states in a single-Cooper-pair box. *Nature*, 398:786, 1999. 15

[79] Daniel Loss and David P. DiVincenzo. Quantum computation with quantum dots. *Physical Review A*, 57:120, 1998. 15, 69

[80] T. Fujisawa, D. G. Austing, Y. Tokura, Y. Hirayama, and S. Tarucha. Allowed and forbidden transitions in artificial hydrogen and helium atoms. *Nature*, 419:278, 2002. 16, 69

[81] Martin Sigrist, Thomas Ihn, Klaus Ensslin, Daniel Loss, Matthias Reinwald, and Werner Wegscheider. Phase coherence in the inelastic cotunneling regime. *Physical Review Letters*, 96:036804, 2006. 16, 70

[82] Jorg Lehmann and Daniel Loss. Cotunneling current through quantum dots with phonon-assisted spin-flip processes. *Physical Review B*, 73:045328, 2006. 17, 70

[83] Sigurdur I. Erlingsson, Yuli V. Nazarov, and Vladimir I. Fal'ko. Nucleus-mediated spin-flip transitions in GaAs quantum dots. *Physical Review B*, 64:195306, 2001. 17, 70

[84] R. Schleser, T. Ihn, E. Ruh, K. Ensslin, M. Tews, D. Pfannkuche, D. C. Driscoll, and A. C. Gossard. Cotunneling-mediated transport through excited states in the Coulomb-blockade regime. *Physical Review Letters*, 94:206805, 2005. 17, 70, 71, 74, 79, 85, 91

[85] Daniel Becker and Daniela Pfannkuche. Coulomb-blocked transport through a quantum dot with spin-split level: Increase of differential conductance peaks by spin relaxation. *Physical Review B*, 77: 205307, 2008. 17, 69

[86] Alex Kamenev and Alex Levchenko. Keldysh technique and nonlinear sigma-model: Basic principles and applications. *Advances In Physics*, 58:197, 2009. 18, 38, 41, 50, 95, 106, 113, 125, 134, 243, 245

[87] L. V. Keldysh. Diagram technique for nonequilibrium processes. *Zh. Eksp. Teor. Fiz.*, 47:1515, 1964. 18, 38, 41
L. V. Keldysh. Diagram technique for nonequilibrium processes. *Sov. Phys. JETP*, 20:1018, 1965.

[88] M. Tews. Electronic structure and transport properties of quantum dots. *Annalen der Physik*, 13:249, 2004. 18, 35, 71

[89] J. K. Furdyna. Diluted magnetic semiconductors. *Journal of Applied Physics*, 64:R29, 1988. 18

[90] Tomasz Dietl. Ferromagnetic semiconductors. *Semiconductor Science and Technology*, 17:377, 2002.

[91] T. Jungwirth, Jairo Sinova, J. Mašek, J. Kučera, and A. H. MacDonald. Theory of ferromagnetic (III,Mn)V semiconductors. *Reviews of Modern Physics*, 78:809, 2006.

[92] R. Beaulac, P. I. Archer, S.T. Ochsenbein, and D. R. Gamelin. Mn^{2+}-doped CdSe quantum dots: New inorganic materials for spin-electronics and spin-photonics. *Advanced Functional Materials*, 18: 3873, 2008. 18

[93] Frederic V. Mikulec, Masaru Kuno, Marina Bennati, Dennis A. Hall, Robert G. Griffin, and Moungi G. Bawendi. Organometallic synthesis and spectroscopic characterization of manganese-doped CdSe nanocrystals. *Journal of the American Chemical Society*, 122:2532, 2000. 18

[94] D. J. Norris, Nan Yao, F. T. Charnock, and T. A. Kennedy. High-quality manganese-doped ZnSe nanocrystals. *Nano Letters*, 1:3, 2001.

[95] Tianhao Ji, Wen-Bin Jian, and Jiye Fang. The first synthesis of $Pb_1 - x Mn_x Se$ nanocrystals. *Journal of the American Chemical Society*, 125:8448, 2003.

[96] Steven C. Erwin, Lijun Zu, Michael I. Haftel, Alexander L. Efros, Thomas A. Kennedy, and David J. Norris. Doping semiconductor nanocrystals. *Nature*, 436:91, 2005. 18

[97] H. Ohno, D. Chiba, F. Matsukura, T. Omiya, E. Abe, T. Dietl, Y. Ohno, and K. Ohtani. Electric-field control of ferromagnetism. *Nature*, 408:944, 2000. 18

[98] D. Chiba, M. Yamanouchi, F. Matsukura, and H. Ohno. Electrical Manipulation of Magnetization Reversal in a Ferromagnetic Semiconductor. *Science*, 301:943, 2003. 18

[99] J. Fernández-Rossier and L. Brey. Ferromagnetism mediated by few electrons in a semimagnetic quantum dot. *Physical Review Letters*, 93:117201, 2004. 19, 222

[100] Fanyao Qu and Pawel Hawrylak. Magnetic exchange interactions in quantum dots containing electrons and magnetic ions. *Physical Review Letters*, 95:217206, 2005.

[101] N. Lebedeva, H. Holmberg, and P. Kuivalainen. Interplay between the exchange and Coulomb interactions in a ferromagnetic semiconductor quantum dot. *Physical Review B*, 77:245308, 2008.

[102] Fanyao Qu and Pawel Hawrylak. Theory of electron mediated Mn-Mn interactions in quantum dots. *Physical Review Letters*, 96:157201, 2006.

[103] Peter Moraczewski. *Electronic and Magnetic Properties of Manganese Doped Quantum Dots*. PhD thesis, Universität Hamburg, 2009. 19, 222

[104] Friedrich Hund. *Linienspektren und periodisches System der Elemente (Struktur der Materie in Einzeldarstellungen Bd.4)*. Springer, 1927. 19

[105] Ramin M. Abolfath, Pawel Hawrylak, and Igor Žutić. Tailoring magnetism in quantum dots. *Physical Review Letters*, 98:207203, 2007. 19

[106] Ramin M. Abolfath, A. G. Petukhov, and Igor Žutić. Piezomagnetic quantum dots. *Physical Review Letters*, 101:207202, 2008. 19

[107] N. Lebedeva and P. Kuivalainen. Magnetotransport through a magnetic semiconductor single electron transistor. *physica status solidi (b)*, 246:1291, 2009. 19

[108] S. Mackowski, T. Gurung, T. A. Nguyen, H. E. Jackson, L. M. Smith, G. Karczewski, and J. Kossut. Optically-induced magnetization of CdMnTe self-assembled quantum dots. *Applied Physics Letters*, 84: 3337, 2004. 19

[109] Alexander O. Govorov and Alexander V. Kalameitsev. Optical properties of a semiconductor quantum dot with a single magnetic impurity: Photoinduced spin orientation. *Physical Review B*, 71:035338, 2005.

[110] S.-J. Cheng and P. Hawrylak. Controlling magnetism of semi-magnetic quantum dots with odd-even exciton numbers. *EPL (Europhysics Letters)*, 81:37005, 2008.

[111] D. E. Reiter, T. Kuhn, and V. M. Axt. All-optical spin manipulation of a single manganese atom in a quantum dot. *Physical Review Letters*, 102:177403, 2009. 19, 20

[112] Igor Zutic and Andre Petukhov. Spintronics: Shedding light on nanomagnets. *Nature Nanotechnology*, 4:623, 2009.

[113] C. Le Gall, R. S. Kolodka, C. L. Cao, H. Boukari, H. Mariette, J. Fernández-Rossier, and L. Besombes. Optical initialization, readout, and dynamics of a Mn spin in a quantum dot. *Physical Review B*, 81:245315, 2010. 20

[114] Stefan T. Ochsenbein, Yong Feng, Kelly M. Whitaker, Ekaterina Badaeva, William K. Liu, Xiaosong Li, and Daniel R. Gamelin. Charge-controlled magnetism in colloidal doped semiconductor nanocrystals. *Nature Nanotechnology*, 4:681, 2009. 19

[115] J. I. Climente, M. Korkusiński, P. Hawrylak, and J. Planelles. Voltage control of the magnetic properties of charged semiconductor quantum dots containing magnetic ions. *Physical Review B*, 71: 125321, 2005. 20

[116] Alexander O. Govorov. Voltage-tunable ferromagnetism in semimagnetic quantum dots with few particles: Magnetic polarons and electrical capacitance. *Physical Review B*, 72:075359, 2005.

[117] Y. Léger, L. Besombes, J. Fernández-Rossier, L. Maingault, and H. Mariette. Electrical control of a single Mn atom in a quantum dot. *Physical Review Letters*, 97:107401, 2006. 20

[118] J. Fernández-Rossier and Ramón Aguado. Single-electron transport in electrically tunable nanomagnets. *Physical Review Letters*, 98: 106805, 2007.

[119] Ronald Hanson and David D. Awschalom. Coherent manipulation of single spins in semiconductors. *Nature*, 453:1043, 2008. 20

[120] S. Weiss, J., M. Thorwart, and R. Egger. Iterative real-time path integral approach to nonequilibrium quantum transport. *Physical Review B*, 77:195316, 2008. 21, 24, 35, 95, 140, 142, 151, 156, 164, 168, 175, 215

[121] Carsten Timm. Tunneling through molecules and quantum dots:

Master-equation approaches. *Physical Review B*, 77:195416, 2008. 21

[122] A. Komnik and A. O. Gogolin. Resonant tunneling between luttinger liquids: A solvable case. *Physical Review Letters*, 90:246403, 2003. 21, 140

[123] Pankaj Mehta and Natan Andrei. Nonequilibrium transport in quantum impurity models: The Bethe ansatz for open systems. *Physical Review Letters*, 96:216802, 2006. 40

[124] Benjamin Doyon. New method for studying steady states in quantum impurity problems: The interacting resonant level model. *Physical Review Letters*, 99:076806, 2007. 21, 40, 140

[125] Herbert Schoeller and Jürgen König. Real-time renormalization group and charge fluctuations in quantum dots. *Physical Review Letters*, 84:3686, 2000. 22, 140

[126] A. Rosch, J. Paaske, J. Kroha, and P. Wölfle. Nonequilibrium transport through a Kondo dot in a magnetic field: Perturbation theory and poor man's scaling. *Physical Review Letters*, 90:076804, 2003. 22

[127] J. Paaske, A. Rosch, and P. Wölfle. Nonequilibrium transport through a Kondo dot in a magnetic field: Perturbation theory. *Physical Review B*, 69:155330, 2004. 140

[128] Stefan Kehrein. Scaling and decoherence in the nonequilibrium Kondo model. *Physical Review Letters*, 95:056602, 2005. 22, 140

[129] Severin G. Jakobs, Volker Meden, and Herbert Schoeller. Nonequilibrium functional renormalization group for interacting quantum systems. *Physical Review Letters*, 99:150603, 2007. 22, 140

[130] R. Gezzi, Th. Pruschke, and V. Meden. Functional renormalization group for nonequilibrium quantum many-body problems. *Physical Review B*, 75:045324, 2007. 22, 140

[131] Michael Weyrauch and Dieter Sibold. Transport through correlated quantum dots using the functional renormalization group. *Physical Review B*, 77:125309, 2008.

[132] H. Schoeller. A perturbative nonequilibrium renormalization group method for dissipative quantum mechanics. *The European Physical Journal - Special Topics*, 168:179, 2009. 22

[133] C. Karrasch, S. Andergassen, M. Pletyukhov, D. Schuricht, L. Borda, V. Meden, and H. Schoeller. Non-equilibrium current and relaxation dynamics of a charge-fluctuating quantum dot. *EPL (Europhysics Letters)*, 90:30003, 2010. 22, 140

[134] Severin G. Jakobs, Mikhail Pletyukhov, and Herbert Schoeller. Nonequilibrium functional renormalization group with frequency-dependent vertex function: A study of the single-impurity Anderson model. *Physical Review B*, 81:195109, 2010. 22

[135] Kenneth G. Wilson. The renormalization group: Critical phenomena and the Kondo problem. *Reviews of Modern Physics*, 47:773, 1975. 22, 94, 140

[136] N. Andrei. Diagonalization of the Kondo Hamiltonian. *Physical Review Letters*, 45:379, 1980. 22, 94

[137] Akira Oguri. Fermi-liquid theory for the Anderson model out of equilibrium. *Physical Review B*, 64:153305, 2001. 22

[138] Robert M. Konik, Hubert Saleur, and Andreas Ludwig. Transport in quantum dots from the integrability of the Anderson model. *Physical Review B*, 66:125304, 2002. 22

[139] Kenneth G. Wilson and J. Kogut. The renormalization group and the [epsilon] expansion. *Physics Reports*, 12:75, 1974. 22, 140
Kenneth G. Wilson. Renormalization group methods. *Advances in Mathematics*, 16:170, 1975.

[140] Ralf Bulla, Theo A. Costi, and Thomas Pruschke. Numerical renormalization group method for quantum impurity systems. *Reviews of Modern Physics*, 80:395, 2008. 22

[141] Frithjof B. Anders. Steady-state currents through nanodevices: A scattering-states numerical renormalization-group approach to open quantum systems. *Physical Review Letters*, 101:066804, 2008. 22, 140
Frithjof B Anders. A numerical renormalization group approach to non-equilibrium green functions for quantum impurity models. *Journal of Physics: Condensed Matter*, 20:195216, 2008.

[142] Frithjof B. Anders and Avraham Schiller. Real-time dynamics in quantum-impurity systems: A time-dependent numerical renormalization-group approach. *Physical Review Letters*, 95: 196801, 2005.
Frithjof B. Anders and Avraham Schiller. Spin precession and real-

time dynamics in the Kondo model: Time-dependent numerical renormalization-group study. *Physical Review B*, 74:245113, 2006.
Robert Peters, Thomas Pruschke, and Frithjof B. Anders. Numerical renormalization group approach to green's functions for quantum impurity models. *Physical Review B*, 74:245114, 2006.
Frithjof B. Anders, Ralf Bulla, and Matthias Vojta. Equilibrium and Nonequilibrium Dynamics of the Sub-Ohmic Spin-Boson Model. *Physical Review Letters*, 98:210402, 2007.

[143] David Roosen, Maarten R. Wegewijs, and Walter Hofstetter. Nonequilibrium dynamics of anisotropic large spins in the Kondo regime: Time-dependent numerical renormalization group analysis. *Physical Review Letters*, 100:087201, 2008. 22

[144] F. Heidrich-Meisner, A. E. Feiguin, and E. Dagotto. Real-time simulations of nonequilibrium transport in the single-impurity Anderson model. *Physical Review B*, 79:235336, 2009. 23

[145] A J Daley, C Kollath, U Schollwöck, and G Vidal. Time-dependent density-matrix renormalization-group using adaptive effective Hilbert spaces. *Journal of Statistical Mechanics: Theory and Experiment*, 2004:P04005, 2004. 23
Steven R. White and Adrian E. Feiguin. Real-time evolution using the density matrix renormalization group. *Physical Review Letters*, 93:076401, 2004.
Peter Schmitteckert. Nonequilibrium electron transport using the density matrix renormalization group method. *Physical Review B*, 70:121302, 2004.

[146] Lothar Mühlbacher and Eran Rabani. Real-time path integral ap-

proach to nonequilibrium many-body quantum systems. *Physical Review Letters*, 100:176403, 2008. 24, 140, 143

[147] T. L. Schmidt, P. Werner, L. Mühlbacher, and A. Komnik. Transient dynamics of the Anderson impurity model out of equilibrium. *Physical Review B*, 78:235110, 2008.

[148] Philipp Werner, Takashi Oka, and Andrew J. Millis. Diagrammatic Monte Carlo simulation of nonequilibrium systems. *Physical Review B*, 79:035320, 2009. 140, 143
Philipp Werner, Takashi Oka, Martin Eckstein, and Andrew J. Millis. Weak-coupling quantum Monte Carlo calculations on the Keldysh contour: Theory and application to the current-voltage characteristics of the Anderson model. *Physical Review B*, 81:035108, 2010.

[149] Marco Schiró and Michele Fabrizio. Real-time diagrammatic Monte Carlo for nonequilibrium quantum transport. *Physical Review B*, 79:153302, 2009. 24, 140, 143

[150] C. Jung, A. Lieder, S. Brener, H. Hafermann, B. Baxevanis, A. Chudnovskiy, A. N. Rubtsov, M. I. Katsnelson, and A. I. Lichtenstein. Dual-Fermion approach to non-equilibrium strongly correlated problems. *ArXiv e-prints*, 1011.3264, 2010. 24

[151] J. E. Han and R. J. Heary. Imaginary-time formulation of steady-state nonequilibrium: Application to strongly correlated transport. *Physical Review Letters*, 99:236808, 2007. 24

[152] Dvira Segal, Andrew J. Millis, and David R. Reichman. Numerically exact path-integral simulation of nonequilibrium quantum transport and dissipation. *Physical Review B*, 82:205323, 2010. 24

[153] A. P. Jauho. *Theory of Transport Properties of Semiconductor Nanostructures*, page 127. Chapman and Hall, Londin, 1998. 29

[154] M. H. Cohen, L. M. Falicov, and J. C. Phillips. Superconductive tunneling. *Physical Review Letters*, 8:316, 1962. 32

[155] J. Bardeen. Tunnelling from a many-particle point of view. *Physical Review Letters*, 6:57, 1961. 33

[156] Christoph Hübner. Spintransport durch magnetisch dotierte Quantenpunkte. Master's thesis, Universität Hamburg, 2010. 35

[157] Gerald D. Mahan. *Many-Particle Physics*. Kluwer Academic Publishers, New York, third edition, 2000. 37

[158] H. Nyquist. Thermal agitation of electric charge in conductors. *Physical Review*, 32:110, 1928. 37

[159] Herbert B. Callen and Theodore A. Welton. Irreversibility and generalized noise. *Physical Review*, 83:34, 1951. 37

[160] H.-P. Breuer and F. Petruccione. *The theory of open quantum systems*. Oxford University Press, 2002. 38, 44, 52

[161] Stephane Attal, Alain Joye, and Claude-Alain Pillet, editors. *Open Quantum Systems I - The Hamiltonian Approach*. Springer, 2006.

[162] Stephane Attal, Alain Joye, and Claude-Alain Pillet, editors. *Open Quantum Systems II - The Markovian Approach*. Springer, 2006. 38, 52, 53

[163] Wolfgang Nolting. *Grundkurs Theoretische Physik 5/1: Quantenmechanik - Grundlagen*. Springer, seventh edition, 2007. 44, 66

[164] G. Lindblad. On the generators of quantum dynamical semigroups. *Communications in Mathematical Physics*, 48:119, 1976. 52, 53

[165] Wolfgang Nolting. *Grundkurs Theoretische Physik 5/2: Quantenmechanik - Methoden und Anwendungen*. Springer, sixth edition, 2007. 57

[166] P. A. M. Dirac. The Quantum Theory of the Emission and Absorption of Radiation. *Proceedings of the Royal Society of London. Series A*, 114:243, 1927. 57

[167] S. De Franceschi, S. Sasaki, J. M. Elzerman, W. G. van der Wiel, S. Tarucha, and L. P. Kouwenhoven. Electron cotunneling in a semiconductor quantum dot. *Physical Review Letters*, 86:878, 2001. 62, 79

[168] D. V. Averin and Yu. V. Nazarov. Virtual electron diffusion during quantum tunneling of the electric charge. *Physical Review Letters*, 65:2446, 1990. 65, 77

[169] Werner Heisenberg. Über den anschaulichen Inhalt der quantentheoretischen Kinematik und Mechanik. *Zeitschrift für Physik*, 43:172, 1927. 66

[170] J. M. Elzerman, R. Hanson, L. H. Willems van Beveren, B. Witkamp, L. M. K. Vandersypen, and L. P. Kouwenhoven. Single-shot readout of an individual electron spin in a quantum dot. *Nature*, 430:431, 2004. 69

[171] M. Kroutvar, Y. Ducommun, D. Heiss, M. Bichler, D. Schuh, G. Abstreiter, and J. J. Finley. Optically programmable electron spin memory using semiconductor quantum dots. *Nature*, 432:81, 2004. 69

Bibliography

[172] Ireneusz Weymann, Jürgen König, Jan Martinek, Jozef Barnaś, and Gerd Schön. Tunnel magnetoresistance of quantum dots coupled to ferromagnetic leads in the sequential and cotunneling regimes. *Physical Review B*, 72:115334, 2005. 71, 76, 79, 80

[173] Axel Thielmann, Matthias H. Hettler, Jürgen König, and Gerd Schön. Cotunneling current and shot noise in quantum dots. *Physical Review Letters*, 95:146806, 2005. 73, 77

[174] Jürgen König, Herbert Schoeller, and Gerd Schön. Zero-bias anomalies and boson-assisted tunneling through quantum dots. *Physical Review Letters*, 76:1715, 1996. 74

[175] Jürgen König, Jörg Schmid, Herbert Schoeller, and Gerd Schön. Resonant tunneling through ultrasmall quantum dots: Zero-bias anomalies, magnetic-field dependence, and boson-assisted transport. *Physical Review B*, 54:16820, 1996.

[176] Jürgen König, Herbert Schoeller, and Gerd Schön. Cotunneling at resonance for the single-electron transistor. *Physical Review Letters*, 78:4482, 1997.

[177] Axel Thielmann, Matthias H. Hettler, Jürgen König, and Gerd Schön. Shot noise in tunneling transport through molecules and quantum dots. *Physical Review B*, 68:115105, 2003.

[178] Alessandro Braggio, Jürgen König, and Rosario Fazio. Full counting statistics in strongly interacting systems: Non-Markovian effects. *Physical Review Letters*, 96:026805, 2006. 74

[179] D. V. Averin and A. A. Odintsov. Macroscopic quantum tunneling

of the electric charge in small tunnel junctions. *Physics Letters A*, 140:251, 1989. 77

[180] L. J. Geerligs, D. V. Averin, and J. E. Mooij. Observation of macroscopic quantum tunneling through the Coulomb energy barrier. *Physical Review Letters*, 65:3037, 1990. 77

[181] Bernhard Wunsch, Matthias Braun, Jürgen König, and Daniela Pfannkuche. Probing level renormalization by sequential transport through double quantum dots. *Physical Review B*, 72:205319, 2005. 78

[182] J. Fransson, O. Eriksson, and I. Sandalov. Many-body approach to spin-dependent transport in quantum dot systems. *Physical Review Letters*, 88:226601, 2002. 78

[183] I. Sandalov and R. G. Nazmitdinov. Shell effects in nonlinear magnetotransport through small quantum dots. *Physical Review B*, 75:075315, 2007. 78

[184] Dietmar Weinmann. *Quantum transport in nanostructures*. Physikalisch-Technische Bundesanstalt, Braunschweig, 1994. 82

[185] J. Bardeen, L. N. Cooper, and J. R. Schrieffer. Microscopic theory of superconductivity. *Physical Review*, 106:162, 1957. 94

[186] John W. Negele and Henri Orland. *Quantum Many-Particle Systems*. Westview Press (Perseus), 1998. 95, 98, 106, 113, 122, 196, 232, 234

[187] J. Schwinger, editor. *Selected Papers on Quantum Electrodynamics*. Dover Publications Inc., 1953. 95

Bibliography

[188] Yoshinori Takahashi and Fumiaki Shibata. Spin coherent state representation in non-equilibrium statistical mechanics. *Journal of the Physical Society of Japan*, 38:656, 1975. 100

[189] P. B. Wiegmann. Superconductivity in strongly correlated electronic systems and confinement versus deconfinement phenomenon. *Physical Review Letters*, 60:821, 1988.

[190] Eduardo Fradkin and Michael Stone. Topological terms in one- and two-dimensional quantum Heisenberg antiferromagnets. *Physical Review B*, 38:7215, 1988. 100

[191] M.N. Kiselev, H. Feldmann, and R. Oppermann. Semi-fermionic representation of SU(N) Hamiltonians. *The European Physical Journal B - Condensed Matter and Complex Systems*, 22:53, 2001. 100

[192] V. N. Popov and S.A. Fedotov. The functional-integration method and diagram technique for spin systems. *Zh. Eksp. Teor. Fiz.*, 94:183, 1988. 100
V. N. Popov and S.A. Fedotov. The functional-integration method and diagram technique for spin systems. *Sov. Phys. JETP*, 67:535, 1988.

[193] Berry Simon. *Functional Integration and Quantum Physics*. American Mathematical Society, second edition, 2004. 103

[194] H. F. Trotter. On the product of semi-groups of operators. *Proceedings of the American Mathematical Society*, 10:545, 1959.

[195] T. Kato. *Topics in Functional Analysis*, chapter Trotter's product formula for an arbitrary pair of self-adjoint contraction semigroups. Academic Press Inc, New York, 1979. 103

[196] Steven R. White. Density matrix formulation for quantum renormalization groups. *Physical Review Letters*, 69:2863, 1992. 140

[197] Peter Schmitteckert and Ferdinand Evers. Exact ground state density-functional theory for impurity models coupled to external reservoirs and transport calculations. *Physical Review Letters*, 100:086401, 2008. 140

[198] Stephan Weiss. *Nonequilibrium quantum transport and confinement effects in interacting nanoscale conductors*. PhD thesis, Universität Düsseldorf, 2008. 140, 164, 168

[199] Jens Eckel. *Non-Markovian effects in open quantum systems*. PhD thesis, Universität Düsseldorf, 2009. 140, 164

[200] M. Thorwart and P. Jung. Dynamical hysteresis in bistable quantum systems. *Physical Review Letters*, 78:2503, 1997. 164, 168

[201] M. Büttiker. Four-terminal phase-coherent conductance. *Physical Review Letters*, 57:1761, 1986. 164

[202] Yigal Meir and Ned S. Wingreen. Landauer formula for the current through an interacting electron region. *Physical Review Letters*, 68:2512, 1992. 164, 181, 219

[203] 192

[204] Aditi Mitra and A. J. Millis. Spin dynamics and violation of the fluctuation dissipation theorem in a nonequilibrium ohmic spin-boson model. *Physical Review B*, 72:121102, 2005. 221

[205] Aditi Mitra and A. J. Millis. Coulomb gas on the Keldysh contour:

Bibliography

Anderson-Yuval-Hamann representation of the nonequilibrium two-level system. *Physical Review B*, 76:085342, 2007. 221

i want morebooks!

Buy your books fast and straightforward online - at one of world's fastest growing online book stores! Environmentally sound due to Print-on-Demand technologies.

Buy your books online at
www.get-morebooks.com

Kaufen Sie Ihre Bücher schnell und unkompliziert online – auf einer der am schnellsten wachsenden Buchhandelsplattformen weltweit! Dank Print-On-Demand umwelt- und ressourcenschonend produziert.

Bücher schneller online kaufen
www.morebooks.de

 VDM Verlagsservicegesellschaft mbH
Heinrich-Böcking-Str. 6-8   Telefon: +49 681 3720 174   info@vdm-vsg.de
D - 66121 Saarbrücken   Telefax: +49 681 3720 1749 www.vdm-vsg.de

Printed by Books on Demand GmbH, Norderstedt / Germany